Lecture Notes in Economics and Mathematical Systems 643

For further volumes:
http://www.springer.com/series/300

Daniela Wiehenbrauk

Collaborative Promotions

Optimizing Retail Supply Chains with Upstream Information Sharing

Dr. Daniela Wiehenbrauk
Stuttgart, Germany
dsw@whu.edu

ISSN 0075-8442
ISBN 978-3-642-13392-3 e-ISBN 978-3-642-13393-0
DOI 10.1007/978-3-642-13393-0
Springer Heidelberg Dordrecht London New York

Library of Congress Control Number: 2010932764

Cover design: WMXDesign GmbH, Heidelberg, Germany

Printed on acid-free paper

Springer is part of Springer Science+Business Media (www.springer.com)

To my husband Christian

Foreword

The thesis by Daniela Wiehenbrauk examines ways to increase promotion efficiency in retailing through sharing of competitive information. The supply chain structure consists of a manufacturer, multiple retailers and customers. Customers make purchase decisions based on the retail prices offered at each point in time. Retailers have to choose a pricing strategy and decide how much to buy to satisfy demand. Manufacturers satisfy retailer orders. Given a lack of information regarding the other retailer's prices, the inventory levels end up being either too much or too little, depending on the realized prices of the competitors in the market. Motivated by discussions with a leading German retailer the thesis examines an innovative approach to improve the supply chain efficiency. The approach calls for the manufacturer to announce a competitive index each period based on shipments to all retailers. The index provides a signal regarding the potential demand level without revealing the particular retailer who will promote. The thesis focuses on analyzing the impact of this index across the supply chain. The timing of events requires retailers to independently choose promotion frequency, decide whether or not they will promote in a period and finally decide on the order quantity for that period. When the competitive index is provided, then the retailer fine tunes the actual promotion price and decides on the order quantity. Since retailers have to choose decisions while anticipating demand, with or without the competition index, mixed strategy equilibria will exist. These mixed strategy equilibria can be interpreted as the share of stores of a specific retail chain that offer a promotion in a period. Finally, customers are modeled as loyal or smart. The smart customers adjust their purchases based on observed prices. The thesis utilizes a unique data set obtained from the manufacturer and from the retailer. The product category is diapers. Application of the model enables estimation of the benefits of provision of the competitive index. The additional information from the competition index reduces inventory and thus reduces supply costs to do a promotion. This leads to an increase in the frequency of promotions. The net impact is a benefit to the retailer and to the customer. Conditions are derived when the manufacturer, retailer and customer may all be better off with the competitive index information. The thesis provides a complete analysis of a novel approach to improving the supply chain. Presentations of this idea in academic conferences recently have always generated intense interest because the idea is both simple to explain yet complex in its impact. The empirical data suggests a concrete estimate of its impact for the diaper category in the highly competitive German supermarket environment. By combining the insights from marketing, supply chain management and feedback from industry, the thesis provides an excellent example of research that is current, novel and has the potential to have significant and lasting impact on the academic field. Moreover, the thesis should be of great interest to practitioners as well, i.e., manufacturers and retailers alike, since collaborative promotions are the best way to mitigate forecast error stemming from deciding only last minute on actual promotion prices.

WHU – Otto Beisheim School of Management *Arnd Huchzermeier*
WHU – Otto Beisheim School of Management *Martin Fassnacht*
Purdue University *Ananth. V. Iyer*

Acknowledgements

My Swabian origin endowed me with a natural interest in promotions and bargain shopping – the cheaper, the better. With this dissertation, I had the opportunity to explore my Swabian heritage in greater depth.

Now the work is ready and I hope it can be instrumental to managers and academics alike when dealing with collaboration and promotions. It was submitted as doctoral thesis at WHU – Otto-Beisheim-Hochschule and summarizes the research that I conducted at the Department of Production Management, chaired by Professor Dr. Arnd Huchzermeier.

I thank my thesis advisors Professor Dr. Arnd Huchzermeier, Professor Dr. Martin Fassnacht from WHU – Otto Beisheim School of Management and Professor Ananth Iyer from the Grannert Krannert School of Management of Purdue University for their guidance throughout the research process. I am especially thankful to Professor Dr. Huchzermeier for the inspiring research environment he creates at the Department of Production Management with international exposure and his support of my visit to Kellogg University.

Furthermore, I thank Axel Hopp from Metro Group for supporting this research with empirical data and his contributions to this dissertation by discussing the applications. Also, I am grateful to "Stiftung Goldener Zuckerhut" for the financial support of my dissertation.

A big thanks goes to my colleagues for making my time at WHU an enjoyable one. Christian Artmann, Rainer Brosch, Stefan Spinler and Christoph van Wickeren were a great support and sparing partners.

I am very grateful to my family for their support. My parents and my brother back my plans without reservations and are always with me. Last, but by no means least, I wish to thank my husband Christian for lovingly supporting me throughout this work and beyond!

Stuttgart
July 2010

Daniela Wiehenbrauk

Contents

List of Figures

List of Tables

List of Symbols and Acronyms

Superscripts and subscripts

$\hat{(\cdot)}$	Denotes the variables of scenario no information sharing
$\check{(\cdot)}$	Denotes the variables of scenario information sharing
$*$	Indicates optimality
t	Period
$i = A, B$	Retailers
k	Case determined by the critical fractile c

Symbols

α_i	Size of the loyal customer segment at retailer i
β_t	Size of the smart customer segment in period t
β_{0t}	Size of the brand switching customer segment in period t
β_{1t}	Size of the stockpiling customer segment in period t
β_2	Size of the store switching customer segment
$c(\cdot)$	Critical fractile
c_o	Overage cost
$c_u(\cdot)$	Underage cost
CI	Competition Index
CPFR	Collaborative Planning, Forecasting, and Replenishment
$d(\cdot)$	Demand
$\delta(\cdot)$	Best response
Δ	Difference
ECR	Efficient Consumer Response
EDLP	Every Day Low Pricing
f_i	Promotion frequency of retailer i
g	Stockout cost
$\Gamma(\cdot)$	Inventory cost
GCI	Global Commerce Initiative

h	Holding cost
HILO	High low pricing
$\Lambda(\cdot)$	Promotion CHANCE
$\Omega(\cdot)$	Competition RISK
$\mathsf{p}_i = p_l, p_h$	Promotion price vector of retailer i
p_h	High promotion price
p_l	Low promotion price
ϕ	Promotion depth
$\pi(\cdot)$	Expected profit
POS	Point of sales
$q(\cdot)$	Order quantity
r	Regular price
ρ	Reservation price
s_i	Price strategy space of retailer i
σ	Mixed strategy space
SU	Statistical Unit
SKU	Stock Keeping Unit
θ	Demand distribution
Θ	Cumulative distribution of demand
$\Upsilon(\cdot)$	Base profit
VICS	Voluntary Interindustry Commerce Solutions
w	Cost of goods sold
WAPE	Weighted Average Percentage Error

Chapter 1
Introduction

Promotions are at the same time beloved and feared by both traditional food retailers and branded goods manufacturers in today's competitive retail environment. While retailers use promotions to attract customers and retain their market position against discounters, the branded good manufacturers see potential to win the price competition against private labels.

Promotions are beloved due to their large, measurable and immediate effect on a brand's sale (Blattberg et al. 1981). Despite only serving the loyal customer segment, promotions allow the retailer to additionally serve smart customers, who switch between retailers and stockpile in order to find the best deal.

Promotions are feared because they come along with the cost of excess inventory in the supply chain. The demand of the smart customers does not only depend on the price of a retailer but also on the prices of the competitors and thereby is highly volatile. Hence, the customer demand in the case of promotions is difficult to forecast and has not yet been mastered, despite few first success stories (Arminger 2003). The associated forecast error of demand during promotion periods increases the inventory required to support a promotion and therefore counteracts the prospected profit gains from increased sales.

The inventory dilemma can best be described in a doom loop as illustrated in Fig. 1.1. For illustration, we assume two competing retailers in a homogenous product market. A retailer deciding to enter into competitive promotions will order inventory to serve the expected demand without knowing the size of the targeted smart customer segment nor the activity of his competitor.

The success of the retailer's promotion heavily depends on his competitor's activities: In case the competitor does not promote, smart customers will purchase at his store and he runs a risk of stockouts, if he underestimates the size of the smart customer segment. This leads to lost sales and moreover to a loss of customer satisfaction. If, on the other hand, the competitor promotes as well, the smart customers will equally split their demand between the two retailers and both end up with overage stock at the end of the period, leading to high inventory cost.

In either case, the retailer faces the risk of additional cost in terms of increasing inventory cost through overage stock or the cost of lost sales and customer satisfaction in the event of stockouts. Further, the retailer gets entangled in the doom loop:

D. Wiehenbrauk, *Collaborative Promotions*, Lecture Notes in Economics and Mathematical Systems 643, DOI 10.1007/978-3-642-13393-0_1,

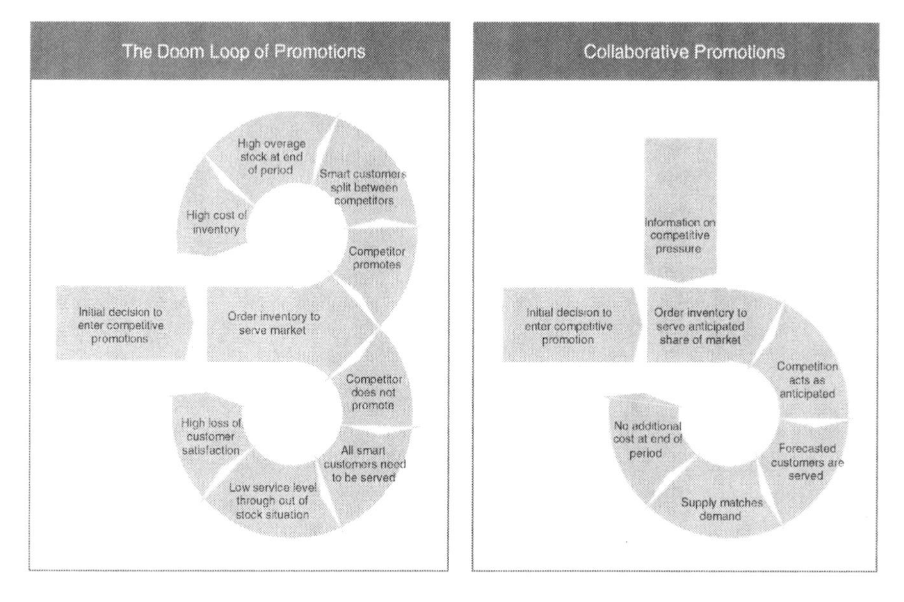

Fig. 1.1 The promotion cycle

Either he is forced to promote again to sell off his excess inventory or in the case of lost customer satisfaction he will try to regain customers by an attractive promotion.

The key question is: How can promotions be elaborated? How can excess inventory on promotions be eliminated in a supply chain, while at the same time providing benefits for customers, retailers and the manufacturer? Is it possible to create a win–win–win situation for all players involved? Retailers and manufacturers will need to find new ways to collaborate on promotions to effectively match supply with demand.

By sharing downstream information from the retailer to the manufacturer on the promotion schedule, manufacturers are able to optimize their production schedules and reduce inventories based on higher market transparency (Iyer and Ye 2000).

An innovative way to employ this information in an upstream manner has been identified by one of Germany's major retailers (Hopp 2005) by compiling a so-called Competition Index. Having access to the promotion schedule of each individual retailer, a manufacturer can aggregate the information on the upcoming competitive pressure in such a Competition Index and in turn share the upstream information from manufacturer to retailer. Based on the Competition Index, each retailer is able to improve his forecast accuracy and adjust the order quantity to avoid excess inventories as well as stockouts.

The former doom loop can therefore be transformed to the Collaborative Promotions loop as shown in Fig. 1.1. Each of the two retailers uses the Competition Index, providing him with an indication of whether his competitor promotes or not. Intuitively, improved knowledge about the promotion activity of the competitor allows an improved synchronization of supply and demand. This in turn eliminates the inventory in the supply chain, making promotions less expensive. Since the cost of a promotion is decreased, the retailer is motivated to increase the frequency of promotions. This leads to lower average prices and thereby increases customer welfare. Moreover, the manufacturer is able to close the price gap to the private labels and to increase his market share.

We are able to show that in a competitive retail environment, collaboration between retailers and manufacturers in the case of promotions by sharing upstream information provides huge potential: inventory in the supply chain is eliminated as orders are more synchronized to actual demand. The supply chain process becomes more efficient, product availability is improved and the customer is served better at a lower price. The supply chain efficiency gains from collaborative promotions are reinvested leading to the desired win–win–win situation for customers, retailers and the manufacturer.

The structure of the subsequent chapters is as follows: Chap. 2 introduces the research problem by giving an overview on the German retail environment, its players, characteristics and current challenges. We define promotion and collaboration, and show how retailers use these tools to compete in their competitive environment. We outline the concept of the Competition Index and the expectations from it. From there, we conclude with the research questions to be addressed in the following chapters.

In Chap. 3, we shall review the literature relevant for the analysis, in order to find answers to the three major questions of this dissertation: How do customers respond to promotions? How do retailers set promotion prices and determine inventory in a competitive environment? And what is the value of upstream information sharing?

The former is closely related to the marketing literature on promotions and customer response, which aims to understand customer behavior in the market. In the second section of this chapter, we focus on the retailer and his decision-making process. We shall show that currently, promotion price and inventory decisions are often taken separately. We discuss promotion price decisions that are based on mixed strategies, as well as those based on inventory levels as determined by the newsvendor model.

Finally, we introduce the literature determining the value of information sharing, where we find various results depending on the type and source of information being shared. While downstream and upstream information is discussed with respect to the type of information, we consider upstream information sharing as the basis for our research.

Chapter 4 contains the formulation of the analytical model. We begin by describing the setup of the model, including the sequence of events, the retailer strategies and moreover the customer demand. We assume three different customer segments: loyal customers and so-called smart customers, which can further be segmented into stockpiling and store-switching customers. Hence, we introduce a model to jointly optimize the retailer's promotion frequency and inventory decision facing this heterogeneous demand. The model is a unique combination of the economic interpretation of promotions as mixed strategies and the operations approach to inventory optimization using the newsvendor model.

The value of sharing the upstream information from the Competition Index is assessed by applying the model to two different scenarios: in the first scenario no information is shared, whereas in the second scenario the retailer receives the Competition Index and can adapt his promotion frequency and order quantity according to the competitive pressure he is about to face. When comparing the two scenarios, it turns out that, with information sharing, supply and demand are better matched, leading to reduced inventory costs for the retailer. We denote this effect as the inventory effect. Further, if promotions come at a lower inventory cost, the retailer can afford to promote more frequently, which is defined as the frequency effect. The inventory and frequency effect add up to a positive profit effect, making the retailer better off with information sharing than with no information sharing. The magnitude of effects, i.e., the profit increase for the retailer and the welfare increase for the customers, therefore depend on the segmentation of the customer base.

As retailers benefit from the customers' willingness to stockpile, so do the customers also benefit as efficiency gains are handed down to the customer in the form of lower prices. The largest benefit is found to be for the store-switching customer segment, followed by loyals and finally the stockpiling customer segment. While still rewarded with a lower price, the stockpiling customers gain least under information sharing, as they already pay the lowest price due, to their willingness to withhold demand when prices are high.

Finally the manufacturer benefits from an increase in sales volume and the resulting gain of the market share, irrespective of the price obtained on the retail level. The desired win–win–win situation can therefore be achieved by means of sharing the Competition Index, but depends on the segment of the customer base and the resulting retailer strategies.

We shall validate this setting based on a unique empirical data set in Chap. 5. Our data set contains two years of sales volumes and prices for a diapers brand for six major German retail chains. We begin by introducing the main characteristics of the retailers, focusing on their retail price format and analyze their promotion strategies.

We propose an appropriate decomposition method to understand customer demand in promotions and fragment the customer base into loyal, stockpiling and store-switching customers. The combination of customer segments each retail chain faces influences their individual promotion strategy.

Based on a second data set containing manufacturer delivery quantities, we compile the Competition Index, and show its power based on its correlation to the competitive pressure of future periods.

Finally, we provide the magnitude of the value of information sharing in a numerical study for two symmetric retailers. The customer segmentation the retailers face, and their own parameters with respect to price levels and promotion frequency, are based on the characteristics of a retailer, as observed in the empirical analysis.

In the last chapter, we summarize our findings from literature, modeling and empirical analysis and conclude with managerial insights.

Chapter 2
Promotions and Collaboration in Retailing

Promotions and collaboration are prevalent tools in the German retail environment which aim at increasing profit margins: while promotions are used to increase sales and market share, collaboration with supply chain partners is intended to increase supply chain efficiency and hence to decrease cost. In this chapter, we begin by providing an overview of the key challenges in the German market situation. The subsequent sections show how retailers employ promotions and collaboration to respond to the highly competitive market environment.

2.1 The German Market Situation

The German retail market is the most important market in Europe and the third largest in the world (Metro Group 2006). It is a highly competitive environment characterized by declining sales volumes, a strong presence of discounters and a high share of private labels. As a response, the sales volume has been declining for most categories between 2002 and 2005. Figure 2.1 provides an overview on the percentage change in sales volumes for four key categories. Only the category sanitary paper increased by 3.9%, whereas the other three categories decreased between 0.6% for food and beverage and 11.8% for diapers.

The decrease of average sales volume is related to the weak German economy: modest economic growth and a high unemployment rate induced low consumer confidence. The share of retail spending as part of private consumption decreased from over 40% in the early 1990s to below 30% in 2005 (Metro Group 2006). Combined with an increasing price awareness following the introduction of the Euro, the result is an even stronger focus on prices by the customers (Harms 2004, p. 23).

Discounters in Germany have a strong market position ever since and are steadily and consistently gaining market share in Germany. Already in 2005, discounters represented 40% of the German market and outgrew the supermarket and hypermarket formats (Metro Group 2006, p. 21). Obviously, price is the dominant factor for the success of discounters. However, this success is built on more than cheap prices. For example, discounters provide lean structures: they are located close to where the customer lives and offer a simple and transparent product choice. Discounters

D. Wiehenbrauk, *Collaborative Promotions*, Lecture Notes in Economics and Mathematical Systems 643, DOI 10.1007/978-3-642-13393-0_2,

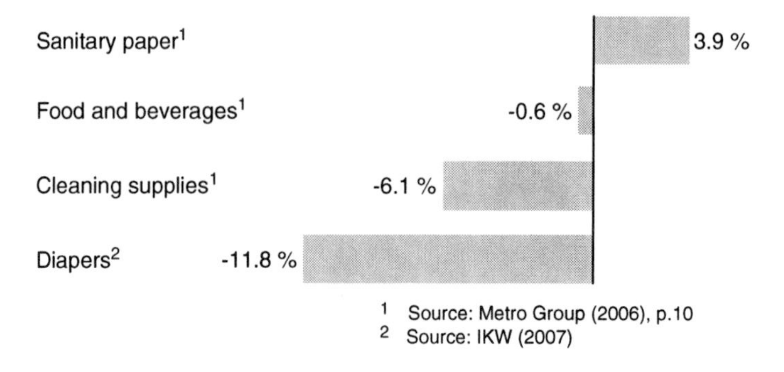

Fig. 2.1 Development of consumer goods categories in Germany: sales volume 2002–2005

thereby enable easy and economical shopping, which is a perfect combination for the budget-conscious customer with no time. Further, discounters have used their brands effectively to overcome the perceived contradiction between low price and high quality, and have succeeded in building trust and a reputation: Aldi is now one of the top brands recalled by German children just behind Lego and Smarties, but ahead of Barbie and Disney (Polman 2004).

It is not only the discounters that are growing but also the private label market. While some years ago private labels were primarily used by branded goods manufacturers to fill up excessive capacity, they have recently become the driver for manufacturers' sales growth. Compared to their branded counterparts, private labels have grown stronger in nearly two thirds of the products studied (Nielsen 2003).

Figure 2.2 depicts the share of discounters and private label brands in Germany in a European context, emphasizing the competitive environment of the German retail market already in 2005. If we consider this in the light of declining sales volumes, it becomes evident that retailers and branded good manufacturers are all chasing a bigger piece of a smaller pie. With the recent market entries of German discounters to other European countries (Schulz 2009), the competitive situation in these countries is comparable to the German situation several years ago.

The question is: how do branded goods manufacturers and retailers compete in this environment? They respond by promotions which puts further pressure on the profit margins. Hence a highly efficient supply chain is required which is achieved by means of collaboration. We shall review both promotion and collaboration and their effects in the following section.

2.2 Promotions

Promotions have become an ubiquitous element of traditional food retailers in order to compete in the German retail environment (Simon and Fassnacht 2009, p. 497). Both retailers and branded good manufacturers use promotions to regain market shares from discounters and private labels.

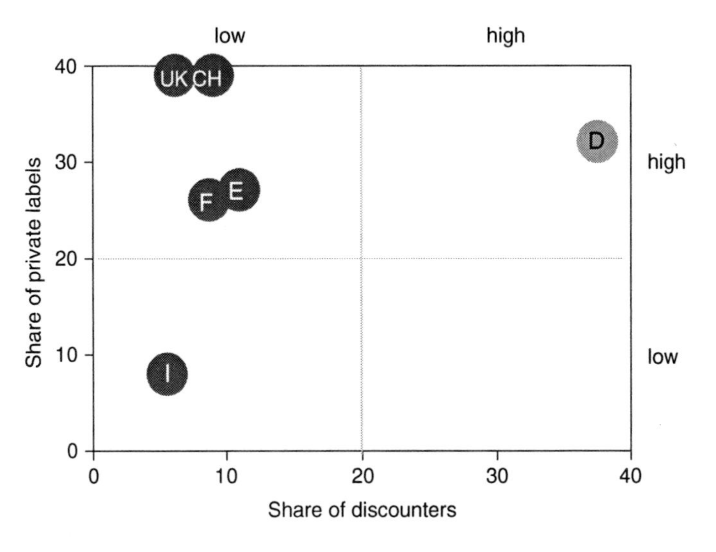

Fig. 2.2 Market share of discounters and private labels in food retailing, 2002 in percent. Source: own illustration based on Metro Group (2006)

Obviously, promotions are an important part of retailers' strategies. We shall first define promotions and consider their impact. Thereafter, different pricing strategies prevalent in the German retail environment are characterized.

2.2.1 *Definition and Impact of Promotions*

Promotional pricing refers to an instance where pricing is the key element of the marketing mix – product, price, place and promotion. Bunn and Banks (2004) define promotions as "a range of tactical marketing techniques designed within a strategic framework to add value in order to achieve specific sales and marketing objectives".

According to Blattberg and Neslin (1990), one can distinguish two major types of promotions. These are firstly customer promotions, which aim at the final customers. Customer promotions can either be offered by the retailer in the form of a price cut, feature or display, or by the manufacturer in the form of couponing or special packs (Simon and Fassnacht 2009, p. 496). The second type is trade promotions, which are offered by the manufacturer to the retailer.

For the purpose of this dissertation, we define promotion as a customer promotion with a reduced price, which is highlighted by advertising and limited in time.

The short term impact of promotions becomes evident in Fig. 2.3: customers respond (Huchzermeier et al. 2002)! The figure displays prices and demand for a diapers brand at a German retailer in 2003. Whenever prices decrease, customer demand increases and can be more than five times higher than demand at regular

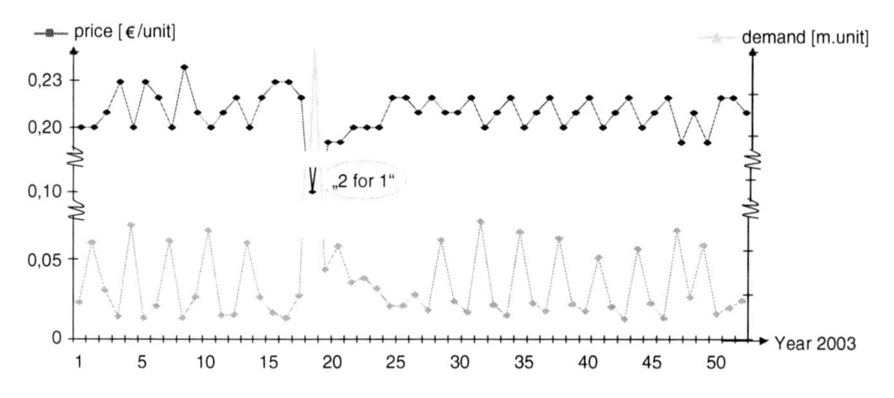

Fig. 2.3 Demand and prices for a diapers brand at a major German retail chain in 2003. Source: own data

prices. However, the volume increase in promotions comes along with high demand volatility: at the same price level, e.g., in week 5 and week 8 in Fig. 2.3, the demand is at 10.2 million units and 8 million units respectively.

With higher volatility of demand, the forecast inaccuracy increases in promotions. A branded goods manufacturer reports an average percentage error (WAPE) of up to 130% for the hygiene category, whereas every-day-business is well under control (Polman 2004). Hence, promotions generate the majority of out-of-stocks, excess inventory and unplanned logistics costs. Not only do retailers and manufacturers lose sales and face higher costs, but moreover, customers become upset if they experience the neglected promise of a promotion in out-of-stock situations and retailers thus put the loyalty of their customers at stake (VICS 2004). In order to improve forecast accuracy in promotions, retailers and manufacturers follow the second trend in the German retail industry: they start to collaborate (GCI and Capgemini 2008).

2.2.2 *Retail Price Formats*

Whether a retailer promotes and, if so, to what extent is characterized by his retail price format. Three retail price formats are prevalent in the German retail environment. These are (1) high-low pricing (HILO), (2) pattern promotions and (3) every day low pricing (EDLP).

HILO retailers alternate prices between a higher regular price and the promotion price, leading to a high variability between the regular and promotion price. They rely on price promotions to attract customers. The HILO retailer often positions himself through providing further customer benefits, such as convenient service, high quality service or a broad assortment of products. According to Tang et al. (2001), the HILO retailer competes on service and assortment, and not on price, in contrast to an EDLP retailer.

While the HILO retailer promotes irregularly, promotions are regular under a pattern promotion strategy. The pattern is characterized by the length of its cycle as well as the sequence of prices within one cycle. Obviously a retailer promoting at a regular pattern is highly predictable to both its customers and its competitors. In terms of customer retention, this is positive: customers adapt to this cycle and the retailer wins them as so called "loyal customers" who purchase in promotions. In terms of competition, the pattern can have negative effects: the retailer is at the risk of being undercut. The impact of pattern promotions has been highlighted by Krishna (1994), but has otherwise received little attention in the literature.

An EDLP strategy differs from a promotional pricing strategy by not emphasizing price specials on individual products, but instead focusing customer attention on good value for money on a regular basis. It involves setting lower average prices with lower variability, i.e., a smaller difference between the regular and promoted price. The EDLP retailer is associated with a wide range of items, but a smaller selection of brands and a less convenient format as compared to the HILO retailer (Bunn and Banks 2004). From a theoretical supply chain perspective, an EDLP strategy provides the advantage of lower demand variability, which in turn reduces stockouts and improves inventory management.

Of course there is a large area of possible combinations in between these three strategies (Lal and Rao 1997). Empirical evidence suggests that the usefulness of the pricing strategy depends on the category. This is because customer price sensitivities can vary by category. In broad terms, the larger the promotional price sensitivity, the more suitable a HILO strategy and the smaller the promotional price sensitivity, the more suitable a strategy with low variation in price, i.e., an EDLP strategy.

Given that promotions are primarily some form of incentive scheme, the extent to which promotions add value is discussed controversially. Sceptics argue that frequent promotions undermine the customer's brand loyalty by encouraging the switching habits of the bargaining customer. However, this is an oversimplification within the highly competitive German retail market, where not to promote would simply result in a loss of market share. A prominent example is Wal-Mart in Germany: refusing to attract the German customer with promotions, Wal-Mart fought eroding market shares and finally backed out of the German market in 2006 (Huchzermeier et al. 2005). Generally, in mature markets, customers view most brands as being adequate for their needs and hence substitutable. They want to be incentivized through promotions to switch between brands.

Also, we should not forget a genuine desire to use promotions as a way to change habitual buying patterns. Promotional objectives can include attracting new or lapsed customers, in other words increasing customer penetration, increasing the loyalty of existing customers or increasing consumption (Simon and Fassnacht 2009).

2.3 Collaboration

Manufacturers and retailers have sought to leverage information sharing and closer collaboration as a way to counter the competitive environment and at the same time improving declining profit margins and reviving growth. This strategy is based on a concept known as Efficient Consumer Response (ECR). ECR dates back to the early 1990s and is a concept which considers that collaboration between retailers and manufacturers is essential in order to "fulfil consumer wishes better, faster and at less cost". Indeed, a widely acknowledged study by Kurt Salmon Associates (1993) projected that greater coordination between supply chain members in the grocery industry could save an estimated $30 billion annually.

ECR is built on four important areas. The first area is *Demand Management*, which comprises all of the considerations associated with understanding and managing the demand for products and services. The second area *Supply Management* focuses on improving the replenishment process of products in the overall supply chain. The third area *Enablers* provides different tools of product identification and data management that are required to permit accurate and timely communication and registration of goods flowing between trading partners. Finally, Collaborative Planning, Forecasting and Replenishment (CPFR) is central to the fourth area of ECR, the *Integrators*.

Given the focus of this dissertation, we focus on the lastly mentioned area of ECR – Collaborative Planning, Forecasting and Replenishment. An introduction to the concept is the subject of the following section. Subsequently, we analyze the pros and cons of CPFR.

2.3.1 Collaborative Planning Forecasting and Replenishment

Collaborative Planning, Forecasting and Replenishment (CPFR) is a business practice that combines the intelligence of trading partners in the planning and fulfilment of customer demand (VICS 2004).

The first CPFR project was initiated in the mid 1990s by Wal-Mart and Warner-Lambert. The partners independently estimated demand six months in advance, compared the forecasts and resolved differences. The pilot provided benefits to both the manufacturer, in terms of a smoother production plan, and the retailer, in terms of a significant reduction of inventory (Seifert 2002). Since then, the framework for CPFR has been refined, resulting in the CPFR reference model.

The CPFR reference model provides a general framework for the collaborative aspects of the planning, forecasting and replenishment processes. Figure 2.4 illustrates this framework in which a retailer and a manufacturer work together to satisfy the demands of the end customer, who is at the center of the model.

The model defines four activities, each including two tasks on both the manufacturer and retailer side. In the first activity, *Strategy and Planning*, the trading partners develop a collaboration agreement and a joint business plan. In the second

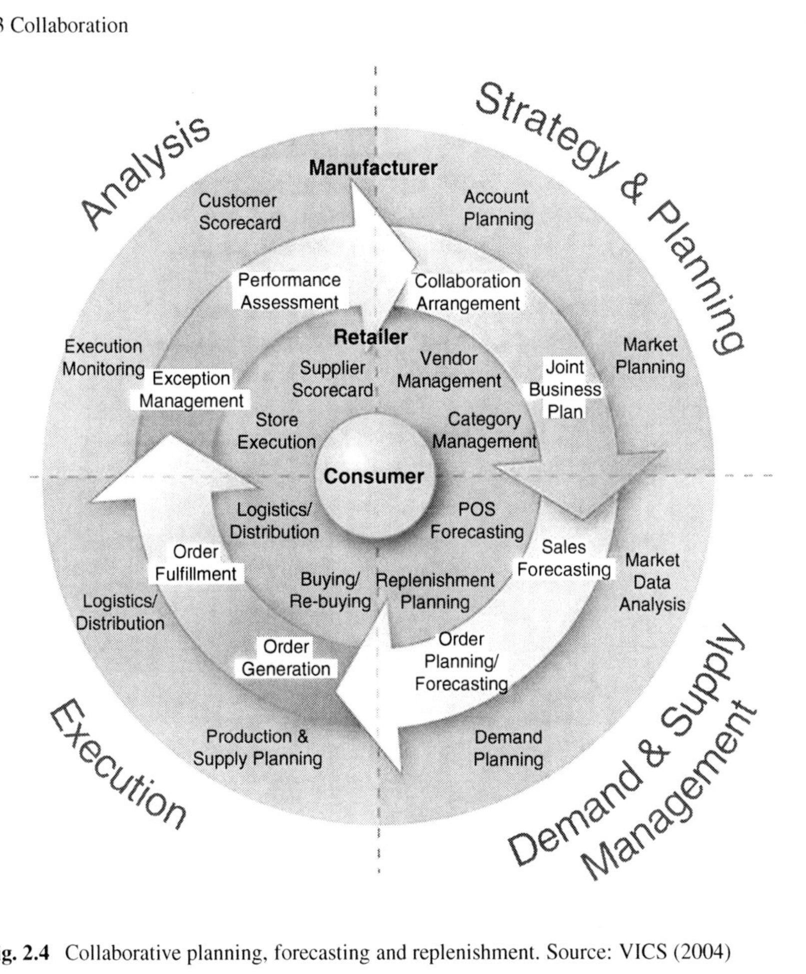

Fig. 2.4 Collaborative planning, forecasting and replenishment. Source: VICS (2004)

activity, *Demand and Supply Management*, the partners collaborate to forecast sales and orders. As supply is required, the order is generated and fulfilled in the third activity, *Execution*. The process concludes with the fourth activity *Analysis*, where exceptions are managed and performance is assessed. Each of the four steps is further broken down to predefined tasks for retailer and manufacturer (VICS 2004).

A recent pilot project at the Metro Group sales division Metro C&C reveals the benefits of CPFR. The CPFR process was designed with a focus purely on promotions rather than for the entire day-to-day business activities. The reason is the higher leverage that was expected from the focus on promotions, as promotions generate the largest swings in demand, and therefore cause high forecast errors. Hence they provide the largest improvement potential.

The pilot project at Metro C&C was set up with seven international suppliers, namely Colgate Palmolive, Henkel, Johnson & Johnson, Kimberly Clark, Lever Fabergé, Procter & Gamble and SCA. The focus was on collaboration in four categories, i.e., detergents, cleaners, hygiene paper and other papers and tissues, with about 4,000 items. The pilot covered over 200 promotion events, lasting from mid 2002 to end 2004. The seven manufacturers formed a test group, while suppliers in the same categories without CPFR integration comprised the control group.

When evaluating the CPFR group as compared to suppliers not in the CPFR group, the following results were reported: sales went up by 25%, whereas in the control group they remained fairly constant at a lower level. The promotional stock service level improved a further percent point and increased to 99.5%. By contrast, the promotional stock service level in the control group declined by 3% to 94.7%. Finally, the CPFR pilot revealed a reduction of the end of promotion coverage: left-over stock levels of the CPFR partners went down by a significant 15% on average to 19 days, whereas the control group remained with 37 days of stock at the end of a promotion (Rode 2005).

What then has been retained from the pilot project? An extension to further categories and retail channels (e.g., at the Real chain) has been prepared but not yet been rolled out. We will describe the major drawbacks of the process in the following section to give some ideas on the hurdles of implementing CPFR.

2.3.2 *Pros and Cons of CPFR*

CPFR has the benefit of combining the inherently unique information of retailers and manufacturers (Aviv 2001). On the one hand, the retailer has information on the point of sales (POS), including regional prices and regional promotions. On the other hand, the manufacturer has information on the overall market perspective, including the competitive pressure in the market. Hence, the benefits reported from CPFR pilots are remarkable: inventory is reduced by 5% to 20% and on shelf availability is increased by 2% to 12% (Metro Group 2006). Also, the pilot of Metro C&C described before, was a success story.

However, there are major obstacles for a thorough rollout of CPFR pilots. The major problem encountered when operating CPFR is the lack of automation. First of all, most systems are island or isolated solutions, which require a different solution for each supplier or retailer respectively. A recent study revealed that over 40% of retailers and manufacturers that engage in CPFR exchange their forecasts manually based on Excel files (ECR 2004).

Secondly, the forecasts of manufacturers and retailers are based on different data sets; the manufacturer forecast is predominantly based on market research data, whereas the retailer forecast builds upon POS data. The market research data are of lower quality and only available after several weeks, but the POS data are available without time lag and at higher reliability. Further, the manufacturer's forecast lacks detailed information such as regional prices and regional promotions, which are available to the retailer. On the other hand, market research data allows getting the "big picture" with cross-regional and cross-brand influences. Obviously, different input data provide different forecast results.

This is also linked to the third obstacle: the collaborative partners use different planning units. The mindset of the manufacturer is in terms of production quantities; for example Procter & Gamble uses Statistical Units (SU) to make a product comparable across different package sizes. In contrast, the retailer forecasts on the level of stock-keeping-units (SKU), which would comprise, for the diapers example, as many as three different package sizes with 40, 80 and 120 units, each representing a different SKU. Consequently, the different planning units have to be made comparable.

Given these drawbacks, a real harmonization of forecasts can be possible for pilot products, but is hardly feasible for complete categories on a day-to-day basis. Instead retailers and manufacturers search for new means of information sharing, which are approached in the next section.

2.3.3 The Competition Index

Retailers identified competitors' price decisions as an important source for the high demand uncertainty in promotions, leading to excess inventory at the end of promotions or stockouts, culminating in high inventory costs.

Information sharing among supply chain partners is crucial to execute promotions efficiently. On the one hand, the retailer has private information about his promotion schedule. On the other hand, the manufacturer has private information about the market perspective. If combined, this information is expected to increase supply chain efficiency.

In order to improve supply chain efficiency, retailers and manufacturers could share the Competition Index, an early indicator on the competitive pressure. The Competition Index is calculated by the manufacturer, based on aggregated information from the retailers: the retailer informs the manufacturer about his planned promotion schedule, the manufacturer aggregates the information across all retailers in the market and derives the so-called Competition Index.

With the promotion schedule, the manufacturer has an important piece of information. No matter how the retailer will actually shape the promotion, the underlying effect remains the same: If a retailer promotes, he faces additional demand from the switching customer segment. Consequently, the order a retailer places will be higher during a promotion than at the regular price. Thus, the manufacturer attains a rough estimate of the retailer's order quantity.

The idea of the Competition Index is based on this simple connection between promotion price and order quantity. The official definition of the Metro Group is the following (Hopp 2005):

Definition 2.1. The Competition Index CI_t reveals the expected competitive pressure of retailer i in the market in period t as a percentage figure. It is calculated as the ratio of expected order quantity q_{jt} in period t across competitors j and the average order quantity across previous periods $t - x, \ldots, t - 1$ and competitors

$j, j \neq i$:

$$CI_{it} = \frac{\sum_j E(q_{jt})}{\sum_j \sum_{t-x}^{t-1} E(q_{jt})}.$$

According to this definition, the Competition Index allows the retailer to conclude whether his competitors are ordering above average, and therefore planning to promote in the upcoming period, or, alternatively, ordering at a level corresponding to demand at regular pricing. The competitors' average order quantity is determined on a rolling average of previous periods, where the number of periods is kept long enough to smooth out seasonalities.

Observe that one cannot reach any conclusions on the ordering or pricing behavior of individual retailers from the above definition of the Competition Index but rather only from the aggregated competition. This has two main reasons. First, the Competition Index is based on a rough estimate of order quantities and is therefore only a description of the competitive pressure in the market. Second, the promotion intensity amongst retailers in the German market environment is high as evident from Fig. 2.5.

The figure shows the price development of five major German retail chains on a weekly basis. Usually, there is more than one retailer offering the product at promotional price. Therefore, extracting a specific retailer is not possible due to the similar market shares of the leading retailers (also see Chap. 6 for a detailed analysis). However, the definition of the Competition Index allows conclusions regarding how the average competitor behaves.

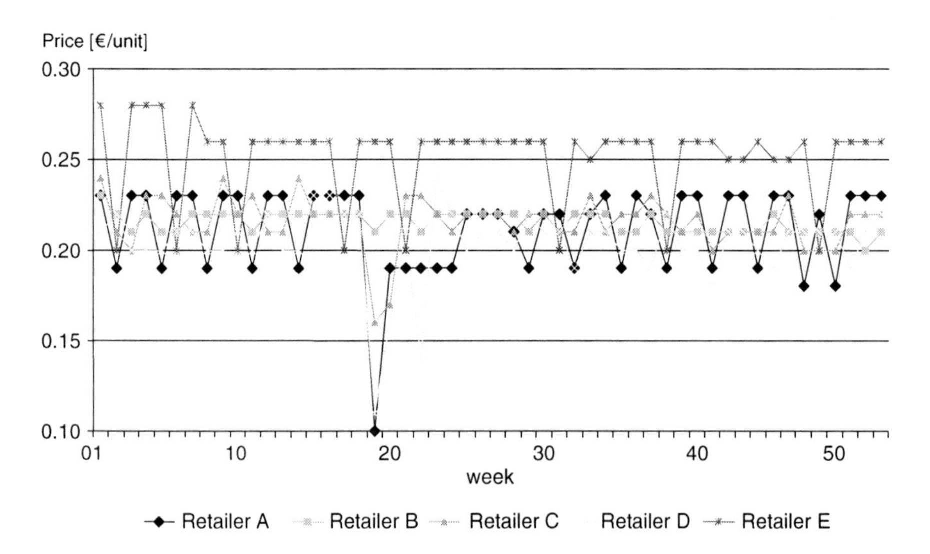

Fig. 2.5 Prices for a diapers brand at five major German retailers, 2003. Source: own data

2.4 Research Questions and Methods

The objective of this dissertation is to determine the value of upstream information sharing, in form of the Competition Index, for customers, retailers and a manufacturer.

The views on the Competition Index are ambivalent. Positive attributes of sharing the Competition Index are: (1) Manufacturers are keen on having early information about retailers' promotions in order to improve production planning (Iyer and Ye 2000). (2) The retailers intention in using the Competition Index is to improve inventory management by reducing out-of-stocks and overstocking. This should not only result in higher payoffs for the retailers but also (3) increase service levels and lower prices for the customers.

However, there are negative views as well: manufacturers expect that with information sharing, retailers would, on average, increase their prices. This in turn is expected to reduce the customer base in the competition with other brands, and thereby decrease the manufacturer's market share. Further, retailers could refrain from sharing their private information about planned promotions due to confidentiality reasons, especially in markets with few players. We have seen earlier that one cannot conclude from the Competition Index who is promoting nor attribute the delivery quantity to a specific retailer in a highly fragmented competitive environment. Finally, competitive regulations could lie in opposition to information sharing, if retailers and manufacturers pool their information in order to maximize their profits by means of reducing competition and increasing retail prices. This behavior would decrease customer welfare and therefore interfere with anti-trust laws.

Given these ambivalent views towards the Competition Index, it is of interest to analyze how the upstream information impacts retailers' and manufacturer's profit as well as customer welfare. We chose a three step procedure to pursue this objective. Figure 2.6 provides a schematic overview:

1. How do customers respond to promotions?
2. How do retailers set promotion prices and inventory in a competitive environment?
3. How does upstream information impact the retailer's decision?

First, we analyze how customers respond to promotions in order to derive the customer demand function. This first step provides an insight as to which customer segments retailers are competing for, and is primarily an empirical task. We shall decompose a unique data set in the diapers category to prove existence and identify the size of three customer segments: loyal and smart customers, where the smart customers can further be segmented into stockpiling customers and store-switching customers.

In order to address the customer segments and gain market share, retailers compete by the use of promotions with a focus on the price sensitive smart customer segments. Whoever offers the lowest price in the market serves the smart customer segments. We shall apply a game theoretical approach and use a mixed strategy

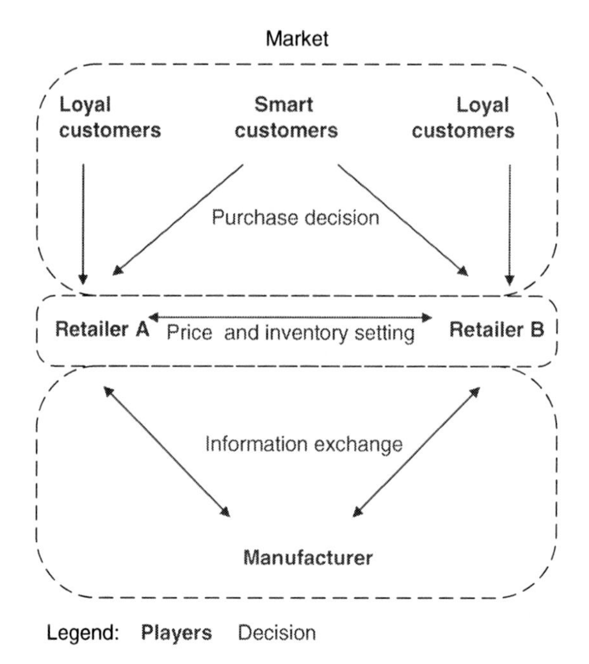

Fig. 2.6 Research overview

equilibrium framework to analyze the price competition between two symmetric retailers, where each retailer determines promotion frequency and promotion depth independently. Further, we shall link this marketing perspective to operations by making inventory an explicit choice variable for the retailer, based on the newsvendor model.

Finally, how does upstream information impact the retailer's promotion price and inventory decision? By comparing the two scenarios of no information sharing and information sharing, we shall see the impact on inventory costs (inventory effect), promotion frequency (frequency effect) and retailer profit (profit effect). Moreover, we shall provide insight into the impact of information sharing on customer welfare as an increasing promotion frequency may lower retail prices on average and hence increase customer welfare.

Chapter 3
Literature Review

The theory underlying our objective to identify the value of upstream information sharing in promotions rests on three pillars. The first pillar encompasses customer demand in promotions, which provides an understanding of how and why customers respond to promotions. The second pillar integrates the retailer's promotion pricing and inventory decisions in a competitive environment. It is a link between the marketing literature on promotions, the operations literature on inventory management and the economics literature on game theory, describing the competition between retailers. Finally, the third pillar determines the value of information sharing between collaborating retailers and a manufacturer, where the value of information sharing depends on the type and source of the information. In the following, we shall introduce each of these three pillars based on the key papers.

3.1 Customer Demand in Promotions

Price promotions are used extensively in marketing for one simple reason: customers respond – at least in the short term (Simon and Fassnacht 2009). In the competitive retail environment with high pressure on margins, it is important to address promotions and the customer response to promotions with sophistication that goes beyond the recognition that promotions increase sales. The decomposition of customer demand in promotions is the starting point of our research and the subject of numerous studies in the marketing literature. It is shown that customers respond to promotions along the three dimensions (1) stockpiling, (2) package size and brand-switching and (3) store-switching.

We shall characterize the three dimensions of customer response subsequently by describing the conceptual background, the underlying theory and by providing empirical evidence. The magnitude of responses is summarized in the last section.

D. Wiehenbrauk, *Collaborative Promotions*, Lecture Notes in Economics and Mathematical Systems 643, DOI 10.1007/978-3-642-13393-0_3,

3.1.1 Stockpiling

Conceptual Background

Stockpiling occurs because promotions induce customers to buy different quantities or at different times than they would have otherwise. Either way, customers end up with a higher quantity than they would have purchased in the absence of a promotion (Ailawadi et al. 2007).

Literature has identified three major consequences of the stockpiling effect. First, stockpiling shifts purchases of the promoted brand forward that would have occurred anyway. Customers stockpile to satisfy future demand at lower cost (Assuncao and Meyer 1993). This "purchase acceleration" reduces the margin for both retailers and manufacturers, given that the profit margin is typically lower during promotions than at the regular price (Blattberg and Neslin 1990).

Second, the additional household inventory suspends the customer from the market and prevents the customer from purchasing different brands or purchasing from different competitors. "Pre-emptive switching" foils the marketing plans of the competitors and is beneficial for retailers and manufacturers (Hendel and Nevo 2004).

Third, stockpiling causes increased consumption induced by the presence of additional household inventory (Bell et al. 2002). The "consumption effect" is beneficial for both retailers and manufacturers. When customers stockpile without increasing their consumption, a temporary expansion in demand is observed, followed by a longer than normal elapsed time before the customer re-enters the market for a subsequent purchase.

Theory

The theory on stockpiling, explaining the phenomenons of "purchase acceleration" and "preemptive switching", is based on two different formal models from the operations literature: the Economic Order Quantity model and the (s, S)-inventory policy.

A model related to the Economic Order Quantity model has been developed by Blattberg et al. (1981). The customer belongs to one of two customer segments, that differ in holding cost and consumption rate. Either customer is assumed to minimize the cost of satisfying his household demand for the product. By buying in promotions, the customer can decrease his purchase cost, but may incur a cost by carrying more inventory of the product. The customer decides on his purchase quantity by trading off holding cost against the reduced price of the item. A similar model is applied by Iyer and Ye (2000). However, the authors allow different reservation prices for each of the two customer segments.

Boizot et al. (2001) determine the customer's optimal purchase strategy for a storable good taking into account price uncertainty. The customer continuously observes the market price and must anticipate randomly arriving price promotions.

In this environment, the optimal purchase policy for the customer is a (s, S)-inventory policy, which implies that, at the beginning of each period, the customer reviews his household inventory level, and if it is below s, he makes a purchase at the retailer to raise his inventory level to S.

Assuncao and Meyer (1993) extend the inventory models by making the consumption amount an explicit decision variable. Their normative model reveals that the consumption rate increases with the size of existing household inventories. That is, given some memory of price, customers will rationally increase consumption when faced with a promotion.

Empirical Evidence

As theory suggests, several researchers have provided empirical evidence that customers respond to promotions with stockpiling.

Blattberg et al. (1981) show that promotions induce "purchase acceleration" and "pre-emptive switching". That is, customers buy larger quantities with longer purchase intervals in promotions than at regular prices. Based on diary panel data, the authors show an increase in duration until the next purchase ranging from 23% to 36% and an increase in the quantity purchased from 8% to 35% for four categories (aluminum foil, facial tissue, liquid detergent and waxed paper). Recent empirical studies show that the stockpiling effect is larger than previously believed. Van Heerde et al. (2002) apply a lagged-demand model to decompose the sales bump in promotions. The authors show that between 26% (tissue) and 44% (peanut butter) of the promotional sales increase can be attributed to stockpiling. Hendel and Nevo (2003) find that households that are larger and reside in suburban locations hold larger volumes of inventory.

Further, literature shows that additional inventory on hand can lead to an endogenous increase in consumption. However, the magnitude of the "consumption effect" is dependent on the category. The results are intuitively plausible: whereas butter and margarine are mostly invariant to the level of inventory on hand, the consumption of hot dogs, ice cream and soft drinks increases with the household inventory (Bell et al. 2002). However, the consumption effect has recently also been reported in seemingly consumption-invariant categories such as laundry detergent and paper towels (Chandon and Wansink 2002) as well as dryer softeners (Bell and Botzug 2004).

Ailawadi et al. (2007) summarize and quantify "purchase acceleration", "preemptive switching" and "consumption effect" for the manufacturer. The authors analyze two categories (ketchup and yogurt) and find that the negative "purchase acceleration" is compensated by the positive impact of the "consumption effect" and "preemptive switching". Further, the authors find an increase in repeated purchases after a promotion. Therefore, the benefits from stockpiling are substantial for the manufacturer.

3.1.2 Brand and Package Size Switching

Conceptual Background

Brand and package size switching describe the customers' decision to "purchase different brands and package sizes from that which would have been purchased had the promotion not been available" (Blattberg and Neslin 1990, p. 112). In this stream of literature, brand-switching is used as synonymous for package size switching: Guadani and Little (1983) note that "different sizes of the same brand are clearly different products. Therefore, we model brand-sizes" (Guadani and Little 1983, p. 213).

Brand switching predominately impacts the manufacturer. For the retailer, brand-switching simply redistributes purchases in the category, possibly at a different margin. Package size switching, however, has further impact on margins, since larger package sizes are commonly priced lower, rewarding the customer with a volume bonus. Brand and package size switching will not be considered in the development of the model, but shall be discussed in this context for the sake of completeness.

Theory

The theoretical explanation as to why promotions induce brand-switching is commonly based on the utility theory (Blattberg and Neslin 1990). A customer is modeled as selecting the brand with the highest utility among those available at the time a choice is made, given his budget constraint. The model most closely related to this theory is the logit choice model (Ben-Akiva and Lerman 1985). Gupta (1988) describes the logit choice model as "appealing, because it is based on behavioral theory of utility, allows explanatory variables and accounts for competition".

Empirical Evidence

Brand switching has been the dominant topic in the marketing literature on promotions, triggered by the findings of Gupta (1988). The author supplements the logit choice model with purchase incidence and purchase quantity models. He finds that brand-switching accounts for 84% of the total sales increase in promotions in the coffee category. The remaining 16% is attributed to stockpiling. Gupta's approach was extended by numerous authors (Chintagunta 1993; Bucklin et al. 1998).

Bell et al. (1999) generalize these findings and confirm them with household-level scanner data on brands of 13 different categories. The authors reconfirm that brand-switching accounts on average for approximately 74% of the sales bump in promotions.

As pointed out by van Heerde et al. (2003), the result cannot be interpreted in a way that if the promoted brand gains 100 units, the other brands in the category lose

74 units. Moreover the authors transform the elasticity results into unit sales effects and conclude that in the promotional week, the other brands lose on average about 33 units, and the category as a whole gains 67 units. This conclusion is based on the same set of 13 categories as in Bell et al. (1999). Consequently, the brand-switching effect is smaller than previously believed, and promotions do not just redistribute purchases in the category, but can actually make categories grow temporarily.

For store level scanner data, Huchzermeier et al. (2002) model a case in which customers react to promotions by package size switching and stockpiling. The authors find evidence that customers behave in a "smart" way by calculating a per unit cost of the product and hence choose package sizes optimally. The logistics regression model estimates customers' decision for a package size to be conditional on the price for each package. For a set of 27 weeks POS-data for 24 stores across Germany on diapers, they report a R^2 ranging from 0.36 to 0.97, with all parameters significant in the t-value. According to the authors, the good accuracy allows the conclusion that customers react in a "smart" way to promotions and choose between package sizes conditional on the price.

3.1.3 Store-Switching

Conceptual Background

Store-switching describes the customer's decision to purchase at a different store than he would have in the absence of a promotion.

Literature identified two different variants of store-switching. The first is "direct store-switching" where the customer's store choice is influenced by outside-store cues (e.g., featured price promotions). Direct store-switching leads to a net decrease in product sales at competing retailers (van Heerde et al. 2004). Further, direct store-switching increases the category sales at the promoting retailer and, moreover, enhances customer traffic which provides the benefit that customers are likely to purchase additional items once they have been attracted to the store (Bucklin and Gupta 1992).

The second is "indirect store-switching" which applies if a household visits multiple stores in a given week. Inside-store cues (e.g., displays with price promotions) influence which items are purchased in the different stores. Indirect store-switching requires cross-shopping of customers, which is common for many households (van Heerde et al. 2004).

Theory

The theoretical explanation of why promotions induce store-switching is commonly based on the utility theory as in the case of brand-switching. Customers assign a utility to a product price and subsequently use this utility to determine their store choice (Cachon and Kök 2007).

Empirical Evidence

Empirical evidence on store-switching is less plentiful than it is on brand-switching, due to the difficulty of attaining an appropriate data set. Further, store-switching seems to depend on the category: Bell and Lattin (1998) provide empirical evidence that customers make their store choice based on the total basket utility. Promotions of a product that is purchased frequently and priced high, such as diapers or coffee, leads to cherry-picking by customers and thus switching between stores (Kumar and Leone 1988; Lal and Rao 1997). Kumar and Leone (1988) find statistically significant price cross-elasticities between stores for promotions of diapers. Also, van Heerde et al. (2002) account about one-third of the sales bump in the tuna category to store-switching.

Richards (2005) has the objective to determine whether supermarket retailers in a specific geographic market use price promotions in order to gain market share from their rivals. Based on a nested logit model of retail pricing, he finds that customers switch between stores if stores are highly substitutable but prices are not. He further finds that the elasticity of substitution among supermarkets is indeed significantly lower than the elasticity of substitution within each store. Therefore, a price promotion is likely to build store traffic, but also cannibalizes sales from non-promoted products.

The remainder of the literature on store-switching focuses on the lasting transition of (loyal) customers from their preferred store. Rhee and Bell (2002) find that a transition is not related to temporary price reductions. However, the authors confirm that customers undoubtedly switch for some trips on the basis of price changes (Rhee and Bell 2002, p. 234). Similar results are described in Leszczyc et al. (2000) and Fox et al. (2004).

3.1.4 Decomposition of Customer Demand in Promotions

Two recent decomposition studies, namely van Heerde et al. (2004) and Ailawadi et al. (2007), summarize the magnitude of customer responses as shown in Table 3.1.

Van Heerde et al. (2004) introduce a method that allows a unit sales effect to be decomposed into its constituent sources. Their estimation is based on store level scanner data instead of household level data, in order to provide managers with a meaningful decomposition of sales effects in promotions. Their standard decomposition divides own brand unit sales effect into brand-switching, stockpiling and category expansion effect. The stockpiling component includes "purchase acceleration" and "pre-emptive switching". The category expansion effect is further decomposed into store-switching and consumption effects. Finally, the decomposition depends on the type of support (display, feature) for a temporary price discount. The results for the categories studied are summarized in Table 3.1.

With their decomposition being based on unit sales, the authors disprove the estimated magnitude in previous decomposition studies. These consistently found

Table 3.1 Decomposition of the response of smart customer segments (percent). Source: Van Heerde et al. (2004) and Ailawadi et al. (2007)

Author	Category	Stockpiling			Brand switching	Store switching	Others
		Purchase acceleration	Preemtive switching	Consumption effect			
van Heerde et al. (2004)	Tuna	38		6	31	25	
	Tissue	35			21		43[a]
	Shampoo	34			31		33[a]
Ailawadi et al. (2007)	Yogurt	4	5	56	34		
	Ketchup	8	9	39	43		

[a] Category expansion effect

that brand-switching is the largest effect with 74% on average, leaving the remaining 26% to stockpiling. However, they did not take into account possible category expansion effects due to promotions, and consequently underestimated the stockpiling component of the customer response.

Ailawadi et al. (2007) focus on the decomposition of promotion-induced customer stockpiling by explicitly taking into account all three components of stockpiling: "purchase acceleration", "pre-emptive switching" and "consumption effect". Further, the authors consider whether the promotion induces a repeated purchase or not. Applying a nested logit framework to household level data, the authors find the results as summarized in Table 3.1 for the product categories under consideration.

The authors find that the benefits for the ketchup and yoghurt category from a promotion are substantial. This is in particular due to the consumption effect. The benefits from an increased consumption offset the negative impacts, that is purchase acceleration by loyal customers who would have purchased the brand at a regular price at a later date anyway.

3.2 Retailer Decisions

In this section, we introduce an integrated model of a retailer who considers promotion and inventory decisions jointly. We combine the demand and supply side of the retailer decision making or, more generally, we combine marketing and production decision making. The problem is how to obtain optimal inventory ordering and promotion decisions jointly, in order to maximize the total profit in the face of competition.

These problems arise in many business environments, where marketing and production management is critical in business decision making. It is necessary to evaluate the trade-off between the benefits from higher sales in promotions and additional inventory costs. On the other hand, a promotion plan must be supported by a coordinated procurement plan to ensure sufficient stock is available to meet the stimulated demand (Cheng and Sethi 1999).

It is obviously desirable to adopt an integrated decision-making system by considering the two types of decisions jointly. However, most marketing literature does not consider inventory, and there is very few production literature focusing on promotions. In the following, we shall review the literature about these two important retailer decisions: promotion pricing and inventory choice. The promotion pricing literature focuses on the incentives for retailers to promote, whereas the inventory literature focuses on executing promotions efficiently.

3.2.1 *Promotion Pricing*

Why are promotions used, rather than simply charging one price? In retail industry, deviations from the "law of one price" are the norm, rather than the exception.

The purpose of this chapter is to offer *economic* explanations for promotions and to derive the optimal promotion strategies for the retailer therefrom. The theoretical literature on price promotions commonly builds on some kind of heterogeneity of the customer base. Retailers then use promotions to price discriminate amongst the customer segments. The rational for this behavior depends on whether the retailer is a monopolist or competes with other retailers.

We begin by reviewing the research pertaining to a monopolist retailer and subsequently go on to look at the research pertaining to competing retailers. We focus on why the retailer offers promotions and on how his optimal promotion strategy is described. Finally, we provide an overview on literature considering promotions, not only as means to exploit customer heterogeneity but, over and above this as competitive response.

3.2.1.1 Monopolist Retailer

A monopolist retailer has an incentive to promote, if customers are heterogenous in their holding cost (Blattberg et al. 1981; Iyer and Ye 2000), price elasticities (Narasimhan 1984) or search cost (Banks and Moorthy 1999).

Stockpiling

Blattberg et al. (1981) argue that promotions shift inventory holding costs from the retailer to the individual customer. In their model, promotions encourage low-holding-cost customers to stockpile, and the retailer thus only holds the inventory for the high-holding-cost customers. If the size of the low-holding-cost customer segment is large enough, it is profitable for the retailer to offer a promotion. Otherwise, the retailer charges all customers the regular price in return for holding the inventory.

Iyer and Ye (2000) further show that if there is high demand uncertainty, it is optimal for the retailer not to promote. An important finding from a supply chain perspective is the link between the customer's inventory holding cost on the one hand, and retailer and manufacturer profits on the other hand: whenever customers' inventory cost decreases, stockpiling increases. This implies an increase in the customer's interpurchase time and it is optimal for retailers to promote less frequently. Less frequent promotions induce the stockpiling customer to purchase more at each promotion. The situation increases the retailer's profit, but the manufacturer profit decreases if he is unaware of the promotion schedule.

Price Elasticity

Narasimhan (1984) considers the use of coupons as vehicles to price discriminate amongst customers with different price elasticity. To enjoy the savings from coupons, the customer incurs some cost. Based on a price theoretic model, the author shows that the coupon users are more price elastic than nonusers, and as the wage rate and opportunity cost of time decreases, coupon usage increases. Coupons can thus be used to reach the price-elastic customer. These implications are confirmed in an empirical study.

Search Cost

Banks and Moorthy (1999) find that promotions allow retailers to price discriminate between customers with heterogeneous search cost: promotional prices are only available when offered and only to those who search for them. In their model, random promotions arise from the need to protect the low-price offers from high reservation price customers who would otherwise make use of the promotion and negate the firm's attempt to price discriminate. It allows the authors to develop a reason for monopolists to promote randomly (mixed strategy). The authors further find that competition would increase the intensity of a promotion.

3.2.1.2 Competing Retailers

When retailers compete for heterogenous customer segments with a homogenous product, a common approach is to describe promotions as the economic counterpart of mixed strategies. Under a mixed strategy, retailers randomize their prices and, therefore, do not post consistently low or high prices. The literature suggests that mixed strategies can persist in markets where customers are heterogenous in search cost (Varian 1980; Banks and Moorthy 1999), loyalty (Narasimhan 1988; Raju et al. 1990; Rao 1991), service preference (Rajiv et al. 2002) or stockpiling and consumption (Salop and Stiglitz 1982; Bell et al. 2002).

We will first review the literature on each of these dimensions of customer heterogeneity focusing on the theoretical background in order to derive the optimal promotion strategies for the retailer. We will conclude with a summary and derive a further economic explanation of promotions.

Search Cost

In markets with imperfect information about prices, one can observe two customer segments: uninformed and informed. The uninformed customer purchases the product at a randomly chosen retailer, whereas the informed customer incurs search cost to make his purchase at the retailer with the lowest price. Varian (1980) considers the optimal pricing strategy for multiple retailers facing the demand of these customer segments. In this setting, no single pair of prices can be an equilibrium, because if one retailer fixes his price, the competitor "steals" the informed customers by setting his price slightly below and thereby increases his profits. Consequently, the retailer has an incentive to randomize prices in order to remain unpredictable to competition. Therefore, the optimal strategy for each company is a mixed strategy.

Based on this result, Varian provides interesting insights into the distribution of promotion prices under a mixed strategy: ideally a retailer would like to discriminate in his pricing and charge informed customers the lowest price (to increase the profit-pie) while charging uninformed customers the highest price (to exploit surplus). Nevertheless, the retailer is obliged to charge all customers the same price which provides a U-shaped density function of prices.

Several authors have attempted to test the prediction of this theory. Among them is Villas-Boas (1995) who finds evidence that prices of coffee and crackers are drawn from the expected density function. He interprets the results as supporting the existence of a mixed strategy equilibrium in the prices of these two products. Lach (2002) studies the dispersion of prices across retailers, as well as its persistence over time. With a 48 months data set on both durables (refrigerators) and consumables (chicken, flour, paper) products he confirms the predictions of Varian's theory and the stability of the equilibrium over time.

Loyalty

Narasimhan (1988) considers a market, where two asymmetric retailers compete on prices for customers who differ in their loyalty: loyal and switching customers. The loyal customer segment only purchases at his preferred store, whereas the switching customer segment is price sensitive and purchases at the store with the lowest price. The intuition for the non-existence of a pure strategy equilibrium is similar to Varian (1980). The ambition to win the switching customers stimulates the retailers to undercut each other's prices. Thus, in order to win the switching segment, retailers have to remain unpredictable for the competitor, and a mixed strategy is the optimal one.

This model allows Narasimhan to make the following suggestions for an optimal pricing strategy for asymmetric retailers: A retailer with fewer loyals (weaker retailer) should promote at a higher frequency than a retailer with more loyals (stronger retailer) in order to increase profits. Thus, weaker retailers can effectively compete against stronger retailers through the use of promotions. Raju et al. (1990) extend the results by showing that the discount offered by the stronger brand is larger than that offered by the weaker brand. Consequently, stronger brands promote less frequently but with a higher discount than weaker brands.

Rao (1991) uses a similar model to Narasimhan, but extends the number of decision variables for the retailer: regular prices, promotional prices and promotional frequencies. He applies this model with the objective of explaining the competition between private labels and national brands. In equilibrium, the national brand promotes randomly to attract private label customers, if the size of the switching segment is large enough. Otherwise, it protects its profits from the loyal customers with the regular price. It is shown that mixed strategy promotions provide a sustainable strategy for the national brand to use against private labels.

Simester (1997) generalized Narasimhan's model to include multiple products and general demand functions. He shows that with the number of switchers, the discount increases. Further, retailers offer a higher discount on complementary products than on products for which they sell a substitute.

Stockpiling and Consumption

Salop and Stiglitz (1982) show that customers' stockpiling can lead to price promotions as an equilibrium outcome in a competitive market. The intuition is that, in any given period, some retailer might offer promotions to generate additional sales from customers who stockpile, whereas others choose to forgo these customers in order to obtain the benefit of higher prices. From their model, the authors predict that an increase in customer holding cost will lead to: (1) an increase in the average price, (2) a reduction in the probability of promotion and (3) a decrease in promotional depth. Bell and Hilber (2006) support all three results with empirical data on bathroom tissues, paper towels and liquid detergents.

Bell et al. (2002) integrate the idea in a model of price competition between symmetric retailers in response to stockpiling and the subsequent "consumption effect". It is shown, that the presence of stockpiling and the consumption effect incentivises firms to randomly offer promotions. The authors show that the consumption effect causes (1) more intense price competition, (2) deeper promotions and (3) an increase in the frequency of promotions as an equilibrium outcome. When combined, these findings imply that price competition is more intense in categories with stronger consumption effects. This, in turn, leads to lower average market prices and to lower equilibrium profits. In an empirical study based on eight product categories, the authors find strong support for each of the three implications of the theoretical model.

Service Preference

In order to relax price competition, retailers increasingly choose to differentiate by in-store quality and service offered (Homburg et al. 2002). The asymmetric store positioning is appreciated in a different way by customers with a low service preference versus customers with a high service preference. Rajiv et al. (2002) show that the equilibrium strategies of the competing retailers follow a mixed strategy. Their model suggests that promotions are motivated by both traffic building and customer-retention considerations. The relative importance of these considerations is related to the retailer's service positioning: whereas the high-service retailer exploits promotions to generate traffic, the customer-retention considerations are more salient for the low-service store. Consequently, their analysis indicates that compared to a low-service store, the high-service store more frequently offers advertised sales, but at lower discounts. Thus, the high-quality store should rely more on the "frequency cue" while the low-quality store should rather focus on the "magnitude cue".

Summary

It becomes evident that for retailers with a homogenous product competing for heterogenous customers, the literature considers promotions as economic counterparts of mixed strategies.

However, there is a controversial discussion about the use of mixed strategies. Cachon and Netessine (2004) argue that it is not clear how a manager would actually implement a mixed strategy. For the authors it seems unreasonable to suggest that a manager would "flip a coin" when choosing his strategy, e.g., determining his production capacity. However in our situation, mixed strategies are meaningful. Firstly, they provide an intuitive economic explanation of promotions and secondly, retailers need to be unpredictable for their competitors – otherwise the competitor undercuts the prices and steals the switching or informed customers (Vives 1999). Further, mixed strategies do not allow the customer to learn about retailers with low prices.

Finally, if, instead, the retailer had a pure strategy, he would persistently sell his product at the lower price. This leads us to another undesired situation, which is commonly used as an explanation for the use of promotions: the "prisoner's dilemma" (Simon 1995; Blattberg and Neslin 1990; Lal et al. 1996). The idea is that if retailer A promotes and retailer B offers a regular price, then the retailer that promotes captures an increased market share at the expense of his competitor. Therefore, his competitor enters the game and offers promotions as well. At some point in time, all retailers offer promotions and none of them is able to abandon this strategy, since he would lose customers. Consequently, all players find themselves in a suboptimal situation, since it would be more profitable for each retailer not to offer promotions at all.

3.2.1.3 Competitive Response

The previous section showed that price promotions are used to remain unpredictable for the competitor and to win the heterogenous customer segment. In this section, we focus on promotions as a response to the competitor's actions in order to find answers to the question of how competitors react to each other's price promotions and advertising tactics.

Recent articles have explored the so-called competitive response in the context of retailing. In particular, they study the long-run impact of a promotion by means of impulse-response functions derived from vector autoregressive (VAR) models (Pauwels 2004).

For example, Nijs et al. (2001) analyze Dutch retail chain data on 560 product categories over a 4-year period in order to identify whether a retailer uses promotions as a response to the competitor's promotions. The authors find that the dominant form of competitive reaction is no reaction. If promotions occur as competitive reactions, they do so more frequently in the short run (33%) rather than in the long run (15%). The authors attribute the low reactivity in the long run to budget constraints, considering the expense of prolonged price wars. Overall, they suggest that the power of price promotions lies primarily in the preservation of the status quo in the category.

Pauwels (2004) and also Steenkamp et al. (2005) support the finding that competitive reactions are not significant. It is interesting to observe that these authors predominately consider the competitive response within a category, and not between retailers.

In contrast to competitive response, Shankar and Bolton (2004) find that competitor factors (e.g., deal frequency, relative price levels) explain a high degree of the variance in retailer price strategy. The authors relate the retailer's pricing strategy with six determinants: (1) competitor, (2) category factors (e.g., storability), (3) chain positioning and size (chain factors, e.g., chain size), (4) store size and assortment (store factors), (5) brand preference (brand factors) and (6) customer factors. In this setting, competitor factors explain the most variance in retailer pricing strategies. They also find that only in the cases of price-promotion coordination and relative brand price do category and chain factors explain any variance in retailer pricing.

3.2.2 Retailer Inventory

The literature commonly approaches the retailer's inventory decision in the case of uncertain demand with the newsvendor problem. Research on the newsvendor problem can be traced back to the economist Edgeworth (1888) and has attracted researchers ever since with its simple but elegant structure as well as its rich managerial insights: In its essential form, a retailer facing random demand for a product must decide on how many units of the product to order to maximize expected profits. If the order exceeds the actual demand, the retailer has to deal with overage cost for excess inventory, whereas he faces stockouts and underage cost in case the ordered quantity is smaller than the actual demand. The optimal solution to this problem is characterized by a balance between the expected cost of overstocking and the expected cost of understocking.

The newsvendor problem has been extended to include different objectives and utility functions, multiple products with substitution, multiple locations and pricing. For a good overview, the reader is referred to Khouja (1999).

We will focus on the integration of inventory and pricing in the newsvendor problem. The need to integrate inventory and pricing was first propagated by Within (1955). Within (1955) adapted the newsvendor model and included price dependency of demand. Both order quantity and price are decision variables for the retailer. Recent studies incorporate more marketing features, such as price (Petruzzi and Dada 1999; Hall et al. 2003; Bernstein and Federgruen 2004), promotions (Sogomonian and Tang 1993; Cheng and Sethi 1999) or advertising (Khouja and Robbins 2003), which are subsequently reviewed in the following. We do not consider the case where the inventory is fixed at the beginning of the period and temporary price reductions keep remaining inventory in line with estimates of remaining demand. Rather, we will consider the case where the retailer places his inventory and pricing decision at the beginning of each period (periodic review model). For extensions, we reference to Elmaghraby and Keskinocak (2003) and Fleischmann et al. (2004).

Pricing and Inventory

The predominant marketing element which is integrated into the newsvendor problem is the price. Petruzzi and Dada (1999) provide an excellent review of existing literature on simultaneous pricing and inventory decisions. Based on the additive and multiplicative demand function, the authors further provide a unified framework to integrate the demand in the single-period newsvendor model.

Hall et al. (2003) extend the approach to a multi-period and multi-product setting. The authors study dynamic pricing and inventory-ordering decisions in a setting where manufacturers offer discounts to retailers, and retailers manage a category of substitutable products rather than managing individual brands independently. The retailer decides on his optimal order quantity and pricing for the category, taking into account a possible trade promotion of the manufacturer. Considering the interdependence of brands (cross-price effects) within the category setting, compared to the single-brand setting, allows retailers to increase profits by 15–50%. Further, the authors provide interesting findings on the pass-through, that is the discount passed on to the customers by retailers: if the cross-price effect for a brand is low, the retail pass-through increases with the brand's own-price effect. On the contrary, when the cross-price effect is high, the retail pass-through decreases with the brand's own price effect.

Further, Bernstein and Federgruen (2004) consider the competition amongst retailers in order to derive a Nash equilibrium. Retailers compete on the two dimensions (1) price and (2) fill rate, where the fill rate is defined as the fraction of demand that can be met from existing inventory. The individual retailer's decision is based on a newsvendor model. The authors analyze the setting for different demand functions and different competition scenarios and find that a pure strategy Nash equilibrium exists across all cases. In addition, they find that retailers with higher fill rates demand a higher price from the customer. Thus, in their model, the customer pays the price for a high service level.

Promotions and Inventory

Sogomonian and Tang (1993) investigate the benefit of coordinating promotion and production decisions for a retailer. The decisions include the timing and level of a promotion as well as the order quantity. In order to evaluate the benefits of coordinated decisions, the authors formulate two problems: a baseline model, where the production and promotion decisions are made separately and an integrated model, where the two decisions are made jointly. A numerical example provides evidence that with the integrated approach, the net profit increases by 12%, which justifies the importance of coordinating promotion and production decisions.

Cheng and Sethi (1999) formulate a more general setting for the joint inventory-promotion decision problem by modeling demand as a Markov decision process. The state variable of the Markov decision process represents the state of demand under the influence of marketing activities and other uncertain environmental factors. The authors identify an inventory threshold level for each demand state that, if exceeded, makes a promotion profitable.

Advertising and Inventory

Khouja and Robbins (2003) apply the single-period newsvendor problem to a case in which advertising leads to increased customer demand. The effect of advertising has two components: first, the effect on the expected demand which is assumed to be diminishing and second, the effect of advertising expenditure on the variance of demand. The authors show that expenditures in advertising, in general, increase the expected profit and lead to an increase in the optimal order quantity. However, the expected profit decreases with an increasing variance of demand. The larger the increase in demand variance due to advertising, the smaller the optimal advertising expenditure and the maximum expected profit.

3.3 Supply Chain Collaboration

An important determinant of a supply chain's efficiency is how its members coordinate their decisions. Sharing information has emerged as one of the most critical practices witnessed by the success of initiatives like ECR or CPFR. These initiatives usually involve the pooling of information available at retailers and manufacturers in order to improve decision making.

The focus of this dissertation is to analyze the impact of upstream information sharing on the retailers promotion and inventory decisions and therefrom derive the effects for the supply chain. It seems intuitive that greater levels of information sharing and collaboration should benefit the companies involved in these integration efforts. A growing body of research in operations management attempts to identify the potential arising from information sharing between retailers and manufacturers. The findings provide mixed results in terms of the value of information sharing. An excellent overview on the results is provided in Chen (2003).

The literature determines the value of information sharing predominately in the context of a supply chain without competition. The relevant papers are reviewed first. Thereafter, we extend the focus on competing retailers and competing supply chains.

3.3.1 *The Value of Information Sharing*

According to Chen (2003), the value of information sharing "refers to the performance improvement as a result of an increase in available information".

The literature on the value of information sharing can be structured by the source of information and the type of information. The source of information describes where the information is generated in the supply chain. It can either be provided by the downstream part of the supply chain, i.e., from the retailer, or from the supply chain's upstream part, i.e., from the manufacturer (Chen 2003).

The type of information shared can be (1) inventory level, (2) sales data, (3) sales forecast and (4) lead time (Lee and Whang 2000). Inherently, sales data can only be shared from the downstream part of the supply chain (retailer) and lead time can only be shared from the upstream of the supply chain (manufacturer). Sharing any of these types of information reduces the uncertainty in the supply chain and is factored into a company's decisions. This results in lower safety stocks and/or higher service levels as compared to the case when no information is shared. The papers are categorized by the type and source of information sharing in Table 3.2.

In the following section, we will consider an important impact of information sharing: the reduction of the bullwhip effect. Thereafter, we will review the literature on downstream information and then proceed with a recent trend observable in the literature: upstream information sharing. We shall describe the setup of the models and characterize the value of information sharing in each setting.

3.3.1.1 Bullwhip Effect

It is important to mention that all four types of information in some way lead to a reduction of an effect known as the bullwhip effect. The bullwhip effect refers to a phenomenon where the orders generated by a stage in the supply chain have a higher volatility than the actual demand at that stage. Lee et al. (1997) identified four key operational factors encouraging the bullwhip effect. These factors

Table 3.2 Type and source of shared information

		Source of information	
		Upstream	Downstream
Type of information	Inventory level	Jain and Moinzadeh (2005), Croson and Donohue (2005)	Cachon and Fisher (2000), Kulp et al. (2004), Croson and Donohue (2005)
	Sales data		Gavirneni et al. (1999), Lee et al. (2000), Iyer and Ye (2000)
	Sales forecast	Aviv (2001)	Aviv (2001), Terwiesch et al. (2005),
	Lead time	Chen and Yu (2005), Steckel et al. (2004)	

include (1) fixed cost in production and ordering which encourages order batching, (2) shortage gaming which encourages phantom orders, (3) price promotions which encourage forward buying and (4) errors in demand signaling, which encourage order adjustments.

In the following, we describe how the different sources lead to a reduction of the bullwhip effect.

According to Lee and Whang (2000), the most commonly shared data between supply chain partners is data on inventory levels. The pursued objective is to optimize the overall supply chain inventory by eliminating redundant safety stock, and, at the same time, ensure the required service levels to prevent stockouts. As the partners have more transparency on the stock level of the supply chain, exaggerated orders that would drive the bullwhip effect can be avoided.

Sharing of historic sales data enriches communication in traditional manufacturer–retailer relationships, where retailers commonly communicate their demand exclusively by placing orders. The analysis of sales data by the manufacturer reduces order volatility and contributes to market transparency, and therefore in turn limits the bullwhip effect.

The bullwhip effect can further be reduced by sharing sales forecast information. Sales forecasts are one of the key drivers for inventory decisions. In any given period, a company determines a set of predictions for the demand in future periods based on its information about the competitive environment. By sharing sales forecast information, the know-how of retailer and manufacturer can be combined and sales forecasts can be improved. The higher the accuracy of these forecasts, the lower the safety stocks required within the supply chain. In practice, the sharing of sales forecasts is implemented in CPFR as described in Sect. 2.3.1.

Finally, sharing lead time information can reduce the order uncertainty faced by retailers. With an increasing uncertainty as to the time when the order will arrive, the retailer builds up higher inventories in order to avoid stockouts. These can be reduced by sharing lead time information.

In an empirical analysis of industry-level data, Cachon et al. (2007) find some support that the bullwhip effect is amplified by price promotions. However, the authors conclude that the bullwhip effect is not observed among retailers and in general not among manufacturers, while finding strong support that seasonality matters: while industries with seasonality tend to smooth production relative to demand, industries without seasonalities tend to amplify production volumes at an underlying smooth demand.

3.3.1.2 Downstream Information

Inventory Level

Cachon and Fisher (2000) provide a model to quantify the value of downstream inventory information on both retailers and manufacturers, finding only limited benefit. The model consists of multiple symmetric retailers, who order an integer number of batches from one manufacturer, who allocates orders and in turn orders from a supplier with ample stock. Demand is assumed to be independent and identical distributed (i.i.d) across retailers and across time. In this setting, it turns out that the retailer's information is most valuable to the manufacturer when the retailer's inventory approaches a level that should trigger the supplier to order additional inventory, which is precisely the time when the retailer submits his order. The additional benefit of having inventory level information beyond order information is therefore low: in a numerical study, the authors show that supply chain costs are lowered by an average of 2.2%.

In a simulation study, Zhao and Xie (2002) show that inventory level information allows both retailers and manufacturers to reduce forecasting errors. The authors conclude that sharing downstream information on order plans further improves the performance of the supply chain in terms of cost.

Kulp et al. (2004) empirically examine the benefits manufacturers receive from downstream information at the inventory level. The authors find only a limited effect on manufacturer profitability. However, among the manufacturers sharing information, the majority succeeds with profit margins above average. The authors conclude that "information sharing appears to be a necessary practice to remain competitive, but not sufficient to earn supernormal profits" (p. 440).

Sales Data

Gavirneni et al. (1999), Lee et al. (2000) and Iyer and Ye (2000) analyze the value of sharing downstream sales data and identify benefits for the manufacturer. The authors in general assume the existence of a perfectly reliable exogenous source of inventory. That is, the manufacturer bears the full cost of guaranteeing reliable supply to the retailer. As a result, information sharing has no direct impact on the retailer because his orders are always, independent of information sharing, received in full.

Gavirneni et al. (1999) examines a two-level supply chain with one retailer and one manufacturer. The retailer faces i.i.d. demand and replenishes his inventory by following an (s, S)-inventory policy. The manufacturer has limited capacity and in case he does not have sufficient inventory to satisfy the retailer's order, a partial shipment is made to the retailer, and the retailer obtains the remainder from an external source. The findings indicate that sharing sales data creates value and that information is valuable only if the system has the flexibility to respond to the information.

Lee et al. (2000) assume that the customer demand at the retailer is positively autocorrelated across periods. The manufacturer bears the full cost of guaranteeing reliable supply to the retailer. In this setting, the authors analyze the benefit of sharing the sales data from retailer to manufacturer. In an analytical and numerical study, Lee et al. (2000) show that the retailer attains no direct benefits from information sharing alone. However, the shared information allows the manufacturer to derive more accurate forecasts of future retailer orders than if that kind of information is not exchanged. Thus, the manufacturer maintains lower inventories and incurs lower holding and shortage costs. The benefits to the manufacturer are greatest when the underlying demand is either highly autocorrelated or highly variable.

In a note commenting on Lee et al. (2000), Raghunathan (2001) argues that these greater effects are due to autocorrelated demand. Under autocorrelated demand, the manufacturer can simply track and use the entire order history. Thereby, the demand forecast variance could be significantly reduced, making a formal information sharing mechanism obsolete for the manufacturer.

The information shared in Iyer and Ye (2000) can be classified in between sales data and sales forecast: they consider the case where information is shared from retailer to manufacturer in retailer's promotion plans. Customer demand occurs randomly at the retailer, and customers stockpile products in order to get a good deal (see Sect. 3.1.1). The retailer defines a promotion plan to maximize expected profits over time. Simultaneously, he chooses his inventory policy based on the newsvendor model. The manufacturer is responsible for maintaining the inventory level at the retailer in order to provide a full service level for all orders. In the absence of information sharing of the exact timing of retailer promotions, this implies the inventory position to be maintained at the high level (the promotion level) to support the 100% retail service level.

Sales Forecast

When it comes to sharing information on sales forecasts, researchers raise their concerns about whether information sharing is truthful. This is especially as Cachon and Lariviere (2001) have shown that retailers have an incentive to inflate forecasts to ensure sufficient supply, in the absence of a contractual obligation for the retailer to purchase what he has forecasted.

Terwiesch et al. (2005) resume the idea and demonstrate the pitfalls of information sharing on the demand forecast in a repeated relationship between one buyer and multiple suppliers in the semiconductor equipment industry. In this industry, buyers provide their suppliers with order forecasts of 24 months. While it seems intuitive that early information sharing should benefit both parties, the practice suffers from two problems. First, the buyer continuously updates his forecasts when being exposed to new information (forecast volatility). Second, a forecast is not legally binding, allowing the buyer to forecast an order without the obligation to actually purchase it (forecast inflation).

Forecast volatility and forecast inflation make it difficult for the supplier to decide when the forecast information is accurate enough to begin production. A supplier that acts immediately on any given forecast is most likely to face significant future rework cost. However, given the long production times in the industry, on-time delivery requires suppliers to start working on an order while it is still a forecast. The extent to which the two parties choose to cooperate, i.e., the buyer shares reliable demand information and the supplier delivers on-time, is analyzed in a multiperiod setting: both parties consider the outcome of previous periods when deciding whether they should cooperate in the current period.

The results show that the suppliers penalize the buyer for unreliable forecasts by delaying the fulfilment of forecasted orders. This is true for both sources of forecast unreliability – forecast volatility and forecast inflation. Vice versa, the buyer penalizes those suppliers that have failed to deliver previous orders on time by providing them with inflated forecasts. The authors conclude that this tit-for-tat behavior of buyer and suppliers is in line with earlier predictions of repeated prisoner dilemma games.

3.3.1.3 Upstream Information

Inventory Level

In Jain and Moinzadeh (2005), the manufacturer provides the retailer with information on the inventory status at the warehouse. In their model, the manufacturer acts as wholesaler to retailers and directly to walk-in customers. While the walk-in customer receives no information, the retailer receives information about product availability at the time he places his order. To take advantage of the information, the retailer changes from a single-level base-stock policy to a two-level, state-dependent base-stock policy. Under this policy, the retailer verifies availability of the stock at the manufacturer when end-customer demand occurs. He then uses two different base-stock levels, each corresponding to whether the product is available at the manufacturer or not.

In a numerical analysis, the authors show that information sharing reduces the average percentage cost at the retailer by 9.5%. However, there is evidence, that retailers cannibalize the information by playing the capacity rationing game: in order to avoid longer lead times, retailers place large orders when the product is not in stock. This abuse of information by the retailer triggers the bullwhip effect in the supply chain and moreover reduces the service level for walk-in customers.

Consequently, upstream information sharing of the inventory level results in a win for the retailer, but a loss for both manufacturer and walk-in customers.

Croson and Donohue (2005) use a controlled version of the Beer Game to examine whether downstream or upstream inventory level information is more effective in reducing the variance of orders, i.e., in reducing the bullwhip effect. The authors find no benefit from upstream inventory information. Only downstream information sharing leads to a significantly lower variance of orders throughout the supply chain. Further, the authors find that upstream supply chain members benefit the most from downstream information sharing.

Sales Forecast

Aviv (2001) studies the case where not only demand realizations, but also the values of other explanatory variables, can be shared in a two-stage supply chain with one retailer and one supplier. In this environment, demand arises stochastically at the retailer, who replenishes his inventory from the supplier, who in turn orders from an outside source with ample stock.

Each member periodically updates its forecast as more information on future demand becomes available. The information includes planned promotions or other relevant external influences (e.g., weather conditions in the beverage industry). This approach used to model the evolution of demand and forecast is contained the Martingale model of forecast evolution (MMFE) framework.

Aviv (2001) applies the model to three different scenarios: (1) the baseline setting, (2) local forecasting and (3) collaborative forecasting (CPFR). In the latter, the supply chain partners jointly maintain and update a single forecasting process in the system.

The comparison of resulting supply chain cost benefits between scenario 1 and scenario 2 reveals the benefits of the integration of forecasting into the replenishment process. The step from scenario 2 to scenario 3 then demonstrates the benefits of collaborative forecasting. Based on a numerical study, the paper finds that compared to the baseline setting, integrating advance demand forecast information into the replenishment process, reduces the supply chain cost on average by 11.1%. With a closer coordination of the supply chain by a collaborative forecasting between retailer and supplier the supply chain costs are additionally reduced by 9.5%. However, the results suggest that it is mainly the diversification of forecasting capabilities that matters; in other words, it is whether or not the trading partners can bring something unique to the table in terms of know-how in forecasting influences.

Lead Time

Another important piece of information is information about the lead-times from the manufacturer to the retailers. Chen and Yu (2005) quantify the value of the upstream information for the retailer. The sequence of events is as follows: At the beginning of each period, the retailer places his order, if any, with the supplier. Within the period, demand arises as i.i.d. variables. Excess inventory at the end of the period incurs holding cost, whereas stockouts cause penalty costs.

The product is supplied within a lead time, where the lead time is modeled as a Markov chain with finite state space. The supplier observes the state of the Markov chain and can either share the information with the retailer or not.

A comparison of the two scenarios in a numerical study shows that the value of lead-time information can be significant. This is especially true when the lead-time distribution underlies high volatility or when the demand is of high volume.

It is important to note that Chen and Yu (2005) focus purely on the result of information sharing between supplier and retailer, without changing the underlying lead time. The effect of a reduced lead time is examined in a simulation study by Steckel et al. (2004): a study that is closely related to the Beer game. They find that reducing the lead time by half results in significant cost reductions across the supply chain.

3.3.2 Information Sharing in a Competitive Environment

In the previous papers the customer demand for each retailer is assumed to be independent, and hence, there is no competition among retailers. Now, we introduce competition and assume that retailers and/or manufacturers have private information. The key question is whether information sharing will lead to an equilibrium outcome in some non-cooperative game or not. Information sharing will emerge as an equilibrium outcome if information sharing provides a higher value than no information sharing.

We begin by considering papers that deal with information sharing among horizontal competitors. Subsequently, we review papers that attempt to generalize the horizontal information sharing literature to vertical information sharing in supply chains.

3.3.2.1 Horizontal Information

In this section, we consider the incentives for companies to share their private information in an oligopoly, i.e., horizontal information among competitors.

An early stream of literature in economic journals has investigated the incentive for companies to share their private information horizontally with competition. This has been pioneered by Novshek and Sonnenschein (1982), followed by Vives (1984), Gal-Or (1985, 1986) and Li (1985) among others.

In the typical model, an oligopoly of companies face a linear demand function with an a priori unknown demand parameter. The companies produce products that are either substitutes or complements and engage in either Cournot or Bertrand competition. Before deciding on its quantity or price, each individual company observes

Table 3.3 Literature overview on horizontal information sharing

		Cournot		Bertrand	
		Substitutes	Complements	Substitutes	Complements
Common parameter	Uncertain demand	Do not share (Li 1985)	Share (Vives 1984)	Share (Vives 1984)	Do not share (Vives 1984)
	Uncertain marginal cost	Share (Li 1985)	Do not share (Gal-Or 1986)	Do not share (Gal-Or 1986)	Share (Gal-Or 1986)
Firm specific parameter	Uncertain capacity constraints		Do not share (Farmer 1994)		

a noisy signal of the true value of the demand parameter. Whether the companies have an incentive to disclose their private information depends on the model specifics (Zhang 2002). The results are summarized in Table 3.3.

Vives (1984) shows that in a Cournot duopoly with product complements or a Bertrand duopoly with product substitutes, information sharing of common parameters increases a company's profit. On the contrary, information sharing decreases a company's profit in Cournot competition with substitute products and Bertrand competition with complements. Further, Gal-Or (1986) establishes that the benefit of information sharing depends on the type of information shared: if the information is about private parameters, i.e., marginal cost, the benefits of information sharing are reversed. The results are supported by the findings of Farmer (1994), who shows that under the private parameter uncertain capacity constraints and product substitutes, information sharing does not always increase expected profits of Cournot duopolists. Li (1985) generalizes the above literature by making weaker distributional assumptions about the random variables.

3.3.2.2 Vertical Information

Recently, there have been several attempts to generalize the findings from horizontal information sharing to vertical information sharing in a supply chain. They aim at revealing incentives for competing retailers to share demand information with a common supplier.

Li (2002) examines the incentives for downstream information sharing in a two-level supply chain with one manufacturer and multiple retailers. The retailers sell substitute products in a Cournot competition. The game is initialized by the retailers' decision on whether to share information with the manufacturer. Upon receiving the demand signals (if any), the manufacturer sets the wholesale price and finally retailers choose their quantities. In this environment, the "direct effect" of information sharing on the payoffs is positive for the manufacturer and negative for the retailer. Consequently, no information sharing is the unique equilibrium. Further, the author identifies a second effect of vertical information sharing: the "indirect effect" or "leakage effect", which occurs due to competing retailers who can infer the information from the actions of the informed parties. The author shows that the leakage effect discourages the retailers from sharing their demand information with the manufacturer while encouraging them to share their cost information.

Zhang (2002) supports the finding that no information sharing about uncertain demand is the unique equilibrium. This result holds for either Cournot or Bertrand competition and substitute or complementary products in the case of duopoly retailers.

Chapter 4
Retailer Competition

Retailers offer promotion prices in order to differentiate from competition in mass markets and to win the smart customer segments. However, customers respond with high demand volatility in promotions, leaving retailers stocked with a lot of waste in the promotion channel. In this chapter, we shall model an environment in which retailers receive information upstream from the manufacturer composed in the so-called Competition Index, aiming to understand how information sharing increases supply chain efficiency, both in terms of pricing and inventory management.

The model combines the retailer's promotion price with his inventory decision. Hence, pricing is not only adapted to the competitive environment, but also dynamic customer demand in promotions is better matched with supply.

The approach is fourfold. First, we provide the setup of the retailer game, by describing the sequence of events, the retailers' strategies and the resulting customer demand. Second, we enter the two-staged retailer competition game, by solving for the retailer's optimal order quantity and the equilibrium promotion frequency. Subsequently, we analyze the retailer's order and frequency decision for two different scenarios: no information sharing and information sharing with the Competition Index. Finally, comparing the two scenarios provides insights into the value of information sharing for customers, retailers and manufacturers. We conclude the chapter with an extension of the results for asymmetric retailers.

4.1 Setup of the Retailer Game

In this section, we describe the environment in which the Competition Index is shared. The timing of information sharing is described in the first section. Then we describe the strategies that are available to the retailer and therefrom derive customer demand. The following details apply for both the scenario of information sharing and no information sharing.

D. Wiehenbrauk, *Collaborative Promotions*, Lecture Notes in Economics and Mathematical Systems 643, DOI 10.1007/978-3-642-13393-0_4,

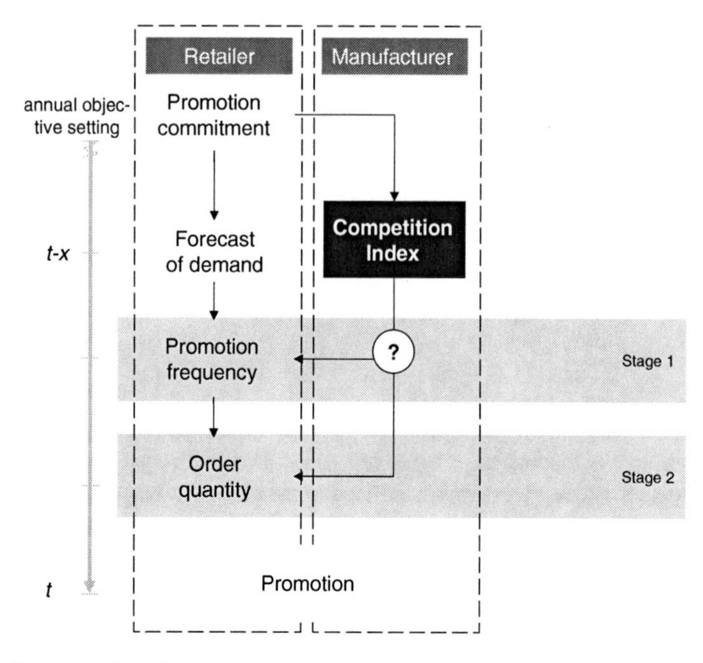

Fig. 4.1 Sequence of events

4.1.1 Sequence of Events

In the initial phase of the retailer competition game, retailer and manufacturer jointly determine the timing of promotions within the annual objective setting, as described in Sect. 2.2. We call this stage promotion commitment as shown in Fig. 4.1. Thereafter, the timing of a promotion is fixed.

Several weeks before the promotion, the retailer builds his forecast of customer demand. Based on his forecast of the customer demand, each retailer has two important decisions to take: promotion frequency and optimal order quantity. We model his decisions in a two-staged game:

In the first stage, the retailers decide on their promotion frequency. Note that we consider retail chains with up to 10,000 individual stores in Germany instead of individual retailers. Hence the promotion frequency does not reflect the number of promotions in a given period. Instead, the promotion frequency specifies at how many stores a retail chain intends to promote in a specific week. If promotion strategies are decentralized, the frequency varies across periods. The optimal promotion frequency is determined as a Nash equilibrium. In the second stage, the retailers determine their optimal order quantity based on a newsvendor model.

In the case of information sharing, the manufacturer aggregates the promotion commitments across retailers in order to determine the Competition Index. He can provide the information to the retailers at two different points in time: either (1) before the retailer sets his promotion frequency or (2) thereafter. In the first case, the retailer exploits the information for both stages of the game, promotion frequency and order quantity. In the second case, the information is only available when determining the optimal order quantity, which in turn influences the decision on promotion frequency.

In order to determine the value of information sharing, we will virtually introduce a stage zero to the game. Both retailers and the manufacturer will review the information sharing agreement in terms of profitability: the retailers directly based on profits and the manufacturer indirectly via market demand and customer welfare. Whenever retailer prices rise, i.e., customer welfare decreases, the manufacturer loses market share to other brands and the manufacturer refrains from sharing information. Only if both players, retailers and the manufacturer, find information sharing to be beneficial, they will agree to share information. We assume truthful information sharing; for a more in-depth explanation see Terwiesch et al. (2005).

4.1.2 Retailer Promotion Strategies

In the first stage of the game, the retailer has to determine his promotion strategy, including both the frequency and the depth of the promotion. His strategy can be considered as complete instructions for the actions to be taken in a two step game: in the first step, the retailer makes a discrete choice as to whether to promote or not at a store. If he has decided for a promotion, he has the discrete choice between offering the product at a low or high promotion price, in the second step.

Formally, we say that in the first step, each of the two retailers i chooses whether he offers a product at promotion price p or regular price r. His pure strategy space s_i in the first stage is $\mathsf{s}_i = \{\mathsf{p}, r\}$. Note that the regular price is the exogenously specified reservation price of the loyal customer segment and the same across all retailers. This is a realistic setting as manufacturers in general provide recommended retail prices. These are equivalent to the customers' reservation price and thus in general not exceeded.

In the second step, the retailers set a concrete promotion price at each individual store, i.e., they decide whether to set the promotion price p at a low level p_l or at a high level p_h, i.e., $\mathsf{p}_i = \{p_l, p_h\}$ with $p_h > p_l$. The pure strategies of retailer i are visualized in Fig. 4.2.

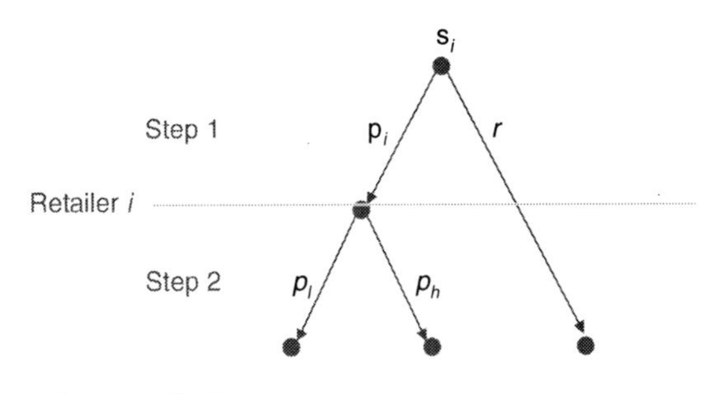

Fig. 4.2 Pure strategies of retailer i

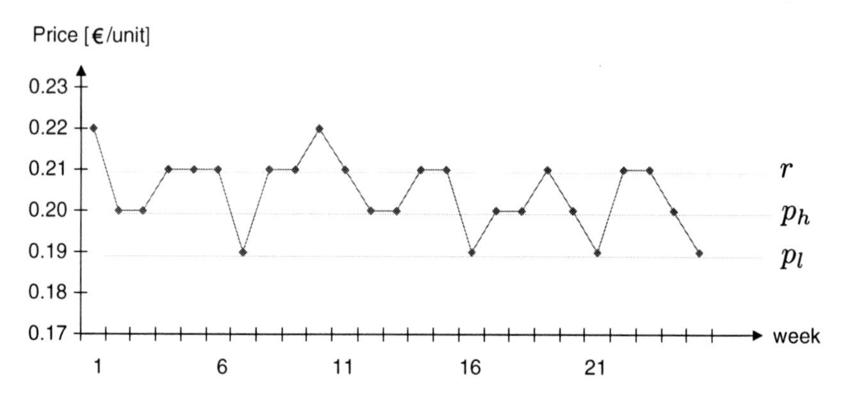

Fig. 4.3 Prices at a German retail chain in 2004. Source: own data

Our setup of having multiple decision variables is similar to that of Banks and Moorthy (1999) and Rao et al. (1995). Further, it is in line with the data set which we shall analyze in the empirical analysis in Chap. 5, where we observe basically three different price levels: one regular price and two different promotion prices, as shown in Fig. 4.3.

So far, we have restricted our attention to pure strategies for both steps. With this restriction an equilibrium may fail to exist (Varian 1980). However, an equilibrium can be found, if retailers choose to randomize their pure strategies, i.e., to play a mixed strategy. With a mixed strategy, a retailer selects a set of prices together with the probability of choosing the different prices. Each player's randomization is stipulated to be statistically independent of those of his opponents.

The literature commonly views mixed strategies as economic counterparts of price promotions as is evident from Sect. 3.2. These allow a meaningful interpretation of the pricing strategies of retail chains (see Chap. 5) with up to 10,000 stores in Germany. Their promotion pricing can be organized in a centralized or a decentralized manner. This means that they may only promote in a certain fraction of their stores, which in itself is a randomization.

Therefore, we provide the following interpretation of mixed strategies for the two steps. In the first step, each individual store of the retail chain decides whether to promote or not. The probability $f_i \in [0, 1]$ then denotes the share of stores that offer a promotion in a specific week of the population "retail chain". The sub-population of promoting stores subsequently sets a concrete promotion price in the second stage, allowing the probability $\phi_i \in [0, 1]$ to denote the frequency with which each promotion price p_l, p_h is charged. We denote f_i as the promotion frequency and ϕ_i as the promotion depth. The closer the promotion frequency is to the bounds of f_i, i.e., 0 or 1, the more centralized is the retailer's promotion strategy.

To be more precise, a promotion frequency $f_i = 0$ implies that all retail stores offer a regular price r, whereas at $f_i = 1$, all retail stores offer a promotion price p. The same logic holds true for the promotion depth, where $\phi = 0$ characterizes a situation in which all promoting stores offer the high promotion price p_h, whereas at $\phi = 1$ all stores offer the low promotion price p_l.

With this interpretation, we follow the suggestion of Vives (1999) that equilibrium points are the outcome of dynamic interactions among rational players (retail stores) in large populations (the retail chain). Further, defining the promotion frequency as the degree to which the retailer offers a centralized vs. a decentralized promotion strategy provides valuable insight into the retail price formats in Chap. 5 and is supported by the available data set.

Given the retailer strategies, we can now determine how customers react to these in the following section.

4.1.3 Customer Demand

Customers are segmented into two types: the loyal and the smart customer segment and each responds differently to the retailer's strategies.

Loyal customers, by definition, buy one unit of the product per period at their preferred retailer i as long as the price of the product is below the reservation price $r < \rho_\alpha$. We denote the loyal customer segment of retailer i by α_i. The loyal customer has adapted to store environments over time. Hence, he derives benefits from his store-specific knowledge so that he is willing to pay higher prices for shopping at retailers that he knows well (Bell and Lattin 1998; Rhee and Bell 2002). We will see in Chap. 5 that the variation in the demand of the loyal customer segment is low, and we can assume a constant size for the loyal customer segment across periods.

The smart customers β_t are bargain shoppers and assumed to be more price sensitive than the loyal customer segment. Their response to promotions can further be segmented into (1) brand-switching, (2) stockpiling and (3) store-switching.

The brand-switching customers β_{0i} are loyal to the retailer i but not to the purchased brand. They switch to the observed product if it is offered at the lowest promotion price. As we consider a one product setup with competition only across retailers but not within categories, we neglect the brand-switching customer segment in our model.

The stockpiling customer β_{1t} is very price sensitive and only enters the market if at least one retailer offers a promotion, and then purchases at the retailer with the lowest price. If the two retailers offer the product at the same price, the demand from the stockpiling segment is split equally. Further, the stockpiling customer balances the cost of holding additional inventory with the potential gains of buying at a lower price. Under this assumption, the size of the segment varies with every period t: the more promotions have occurred in previous periods, the higher the customers' household inventory and the fewer stockpiling customers will be purchasing in the following periods.

Finally, the store-switching customer β_2 purchases on a weekly basis as the loyal customer does. However, he has no preferred retailer and purchases wherever he encounters the lowest price. If the retailers charge the same price it can be considered a tie and the demand of the store switchers is split equally. The store-switching customer segment shows moderate variation across periods and can be assumed to be of constant size in the long term, as shown in Chap. 5.

Table 4.1 Demand matrix for retailer A

			Retailer B		
			$f_B\phi$	$f_B(1-\phi_B)$	$(1-f_B)$
			p_l	p_h	r
	$f_A\phi_A$	p_l	$\alpha_A + \frac{\beta_{1t}+\beta_2}{2}$	$\alpha_A + \beta_{1t} + \beta_2$	$\alpha_A + \beta_{1t} + \beta_2$
Retailer A	$f_A(1-\phi_A)$	p_h	α_A	$\alpha_A + \frac{\beta_{1t}+\beta_2}{2}$	$\alpha_A + \beta_{1t} + \beta_2$
	$(1-f_A)$	r	α_A	α_A	$\alpha_A + \frac{\beta_2}{2}$

With these assumptions, the smart customer segment β_t for our purposes is a composition of stockpiling customers β_{1t} and store-switching customers β_2, i.e., $\beta_t = \beta_{1t} + \beta_2$.

A similar segmentation is applied in Chen et al. (2001) and in Tang and Yin (2007). All three smart customer segments are very well documented in the literature as evident from Sect. 3.1, in terms of both their response to promotions and corresponding equilibrium pricing strategies for the retailers.

Combining the customer segments with the retailers pure strategy set $\mathbf{s} = \{p_l, p_h, r\}$ and the promotion frequency f_i and depth ϕ_i, which are described in the previous section, we can draw the demand matrix for the loyal customer segment and the smart customer segments as shown in Table 4.1.

From the demand matrix, we find the demand for retailer $i = A$ for each of his three price levels $\mathbf{s}_A = \{p_l, p_h, r\}$ as

$$d_A(\mathbf{s}) = \begin{pmatrix} d_A(p_l) \\ d_A(p_h) \\ d_A(r) \end{pmatrix} = \begin{pmatrix} \alpha_A + (\beta_{1t} + \beta_2)\left(1 - \frac{f_B\phi_B}{2}\right) \\ \alpha_A + (\beta_{1t} + \beta_2)(1 - \frac{f_B}{2}(1 + \phi_B)) \\ \alpha_A + \frac{\beta_2}{2}(1 - f_B) \end{pmatrix}. \tag{4.1}$$

For ease of notation, we omit the index t in the demand, but keep in mind that with the weekly variations of the size of the stockpiling segment, demand is time dependent. The above functional form has frequently been considered in the literature. Varian (1980) describes the loyal and smart customer segment as uninformed and informed customers, implying that stores advertise their promotion prices. These are noticed by the informed customers and ignored by the uninformed ones. Narasimhan (1988) and Rao (1991) use a model similar to Varian's, but rather than informed and uninformed customers, they use the expressions "loyals" and "switchers". Huchzermeier et al. (2002) introduce the terminology "smart" customers to describe the behavior of the switching customer segment.

Moreover, the functional form described reflects actual market demand for Pampers Baby Dry as we shall show in Chap. 5. Observe that in the above functional form, we neglect package size. Given that we assume customers to purchase one unit of a product, there is no need to integrate the package size choice. For extensions see Huchzermeier et al. (2002).

4.2 Optimizing the Retailer's Decisions

With the setup in place, this section addresses the retailers' optimization problems: the determination of the equilibrium promotion frequency and the optimal order quantity. These decisions are based on the size of the smart customer segments and necessary in the face of the uncertainty regarding the competitor's pricing.

In the following section, we will describe the optimization problems in a general way which is relevant for both scenarios of information sharing. We begin by solving the second stage of the problem, the optimal order quantity. Then, we derive the retailer's profit function under mixed strategies in order to determine the equilibrium promotion frequency.

4.2.1 Newsvendor Problem: The Optimal Order Quantity

A retailer faces random demand from the smart customer segments given the uncertainty regarding competitor's pricing. Whenever he succeeds in pricing below his competitor, he serves the smart customer segments, otherwise his sales are restricted to the loyal segment or half the respective smart customer segments at a tie.

In this situation, retailer A has to determine his optimal order quantity for each of his three price levels $\mathsf{s}_A = \{p_l, p_h, r\}$ individually to maximize expected profits. The optimal solution to this problem is characterized by a balance between the expected cost of understocking and the expected cost of overstocking and commonly known as the newsvendor problem.

The model formulation presented here is based on Nahmias (2001, p. 250 ff.), Petruzzi and Dada (1999) and Silver et al. (1998, p. 387 f.). The latter show that the cost minimization approach that is usually chosen in standard textbooks is equivalent to the profit maximization approach pursued here.

For each individual price of his pure strategy space $\mathsf{s}_A = \{p_l, p_h, r\}$, the retailer orders $q_A(\mathsf{s}_A)$ units at the beginning of a period before demand is observed. He attains the product at a cost of $wq_A(\mathsf{s}_A)$. If the demand $d_A(\mathsf{s}_A)$ during the period remains below the ordered quantity, then the sales revenue is $\mathsf{s}_A d_A(\mathsf{s}_A)$ and the $q_A(\mathsf{s}_A) - d_A(\mathsf{s}_A)$ units left on hand are inventoried at the unit holding cost h. The total cost for every unit unsold at the end of the period is $h + w$. This cost is called overage cost c_o.

Alternatively, if the demand exceeds the ordered quantity $q_A(\mathsf{s}_A)$, the sales revenue is $\mathsf{s}_A q_A(\mathsf{s}_A)$ and each of the $d_A(\mathsf{s}_A) - q_A(\mathsf{s}_A)$ units short are assessed with the unit penalty cost g. Including the lost margin opportunity, the retailer has the cost of $\mathsf{s}_A + g - w$ for every unit of unsatisfied demand. This cost is called the underage cost $c_u(\mathsf{s}_A)$. Assume that parameters satisfy $0 < w < \mathsf{s}_A$ to avoid trivial solutions. Further, the bargain mentality of the smart customer segment makes them sensitive to underage situations in promotions. Therefore, we assume the unit penalty cost g to be large, i.e., $g > p_h$. Further, the wholesale cost w, holding cost h and penalty cost g are the same for both retailers.

Retailer A's profit for the period $\pi_A(q_A(\mathrm{s}_A), \mathrm{s}_A)$ is then the difference between sales revenue and the sum of holding and stockout cost:

$$\pi_A(q_A(\mathrm{s}_A), \mathrm{s}_A) = \mathrm{s}_A \min\big[q_A(\mathrm{s}_A), d_A(\mathrm{s}_A)\big] - wq_A(\mathrm{s}_A) \\ - h \max\big[q_A(\mathrm{s}_A) - d_A(\mathrm{s}_A), 0\big] - g \max\big[d_A(\mathrm{s}_A) - q_A(\mathrm{s}_A), 0\big].$$

A convenient expression for this profit function is obtained by substituting $c_o = w + h$ and $c_u(s_A) = \mathrm{s}_A + g - w$ and, consistent with Rudi and Pyke (2000, p. 172), using the relationships $\min[q_A(\mathrm{s}_A), d_A(\mathrm{s}_A)] = d_A(\mathrm{s}_A) - \max[d_A(\mathrm{s}_A) - q_A(\mathrm{s}_A), 0]$ and $q_A(\mathrm{s}_A) = d(\mathrm{s}_A) - \max[d_A(\mathrm{s}_A) - q_A(\mathrm{s}_A), 0] + \max[q_A(\mathrm{s}_A) - d(\mathrm{s}_A), 0]$. The profit function can be rearranged as

$$\pi_A(q_A(\mathrm{s}_A), \mathrm{s}_A) = (\mathrm{s}_A - w)d_A(\mathrm{s}_A) - c_u(\mathrm{s}_A) \max\big[d_A(\mathrm{s}_A) - q_A(\mathrm{s}_A), 0\big] \\ - c_o \max\big[q_A(\mathrm{s}_A) - d_A(\mathrm{s}_A), 0\big]. \tag{4.2}$$

Further, let $E(\pi_A) = P_A$ denote the expected value for the profit, $\theta\big(d_A(\mathrm{s}_A)\big)$ the density function and $\Theta\big(d_A(\mathrm{s}_A)\big)$ the cumulative density function of demand, then we obtain

$$P_A(q(\mathrm{s}_A), \mathrm{s}_A) = (\mathrm{s}_A - w)\mu - c_o \int_0^{q_A(\mathrm{s}_A)} (q_A(\mathrm{s}_A) - d_A(\mathrm{s}_A))\theta\big(d_A(\mathrm{s}_A)\big)\partial d_A(\mathrm{s}_A) \\ - c_u \int_{q_A(\mathrm{s}_A)}^{\infty} (d_A(\mathrm{s}_A) - q_A(\mathrm{s}_A))\theta\big(d_A(\mathrm{s}_A)\big)\partial d_A(\mathrm{s}_A).$$

The optimal order quantity that maximizes expected profits of retailer A, denoted as $q_A^*(\mathrm{s}_A)$

$$\max_{q_A(\mathrm{s}_A)} P_A(q_A(\mathrm{s}_A), \mathrm{s}_A) \quad \rightarrow \quad q_A^*(\mathrm{s}_A)$$

is found by solving the first order condition

$$\frac{\partial P_A(q_A(\mathrm{s}_A), \mathrm{s}_A)}{\partial q_A(\mathrm{s}_A)} = -c_o\Theta\big(q_A(\mathrm{s}_A)\big) + c_u(\mathrm{s}_A)\left(1 - \Theta\big(d_A(\mathrm{s}_A)\big)\right) = 0$$

and defined as the critical fractile

$$c(\mathrm{s}_A) = \Theta\big(q_A^*(\mathrm{s}_A)\big) = \frac{c_u(\mathrm{s}_A)}{c_u(\mathrm{s}_A) + c_o} = \frac{\mathrm{s}_A + g - w}{\mathrm{s}_A + g + h}. \tag{4.3}$$

Since $c_u(\mathrm{s}_A)$ and c_o are, given the above restriction, positive numbers, the critical fractile is always between 0 and 1.

To facilitate intuition, the procedure for finding the optimal solution for the newsvendor problem under discrete demand is to determine the value of $q_A(\mathrm{s}_A)$ that makes the cumulative density function of demand $\Theta\big(d_A(\mathrm{s}_A)\big)$ equal to the critical fractile $c(\mathrm{s}_A)$. Consequently, the optimal order quantity for retailer A depends on the opponent's promotion frequency f_B and balances the risk of overstocking

Table 4.2 Optimal order quantities and cumulative densities for retailer A

Promotion price			Regular price	
$q_A^*(\mathsf{p}_A)$	$\Theta_A(d_A(p_l))$	$\Theta_A(d_A(p_h))$	$q_A^*(r)$	$\Theta_A(d_A(r))$
α_A	0	$f_B\phi_B$	α_A	f_B
$\alpha_A + \frac{\beta_{1t}+\beta_2}{2}$	$f_B\phi_B$	f_B	$\alpha_A + \frac{\beta_2}{2}$	1
$\alpha_A + \beta_{1t} + \beta_2$	1	1		

and understocking as characterized by the critical fractile $c(\mathsf{s}_A)$. Retailer A's own promotion frequency $f_A \in [0, 1]$ does not impact his optimal order quantity at this stage.

Table 4.2 visualizes the three discrete order levels in case the promotion price is offered, i.e., $\alpha_A, \alpha_A + \frac{\beta_{1t}+\beta_2}{2}$ and $\alpha_A + \beta_{1t} + \beta_2$ and the two discrete order levels α_A and $\alpha_A + \frac{\beta_2}{2}$ at the regular price, together with the probabilities as given by the demand matrix in Table 4.1, for each of retailer A's pure strategy $\mathsf{s}_A = \{p_l, p_h, r\}$.

In case the critical fractile is between two discrete order levels, it is optimal to choose the order quantity $q_A^*(\mathsf{s}_A)$ corresponding to the higher value. For example, at the low promotion price p_l, the optimal order quantity for retailer A is $\alpha_A + \frac{\beta_{1t}+\beta_2}{2}$ if the critical ratio $c(p_l)$ is $0 \leq c(p_l) < f_B\phi_B$. On the contrary, whenever the critical fractile is $c(p_l) \geq f_B\phi_B$, it is optimal for retailer A to order $\alpha_A + \beta_{1t} + \beta_2$.

If the opponent promotes frequently, there is a low probability for retailer A to gain the smart customers β_t. Then it is optimal for retailer A to reduce his order quantity in order to maximize profits. Whenever the competitor's promotion frequency f_B falls below the critical fractile $c(p_l)$, it is optimal for retailer A to increase his order quantity in order to satisfy an expected increase of demand from the smart customer segment at his stores in case of a promotion.

In order to express the dependency of the optimal order quantity on opponent's promotion frequency, we solve the conditions for the opponent's promotion frequency f_B. For example, the condition $c(p_l) \geq f_B\phi_B$ is solved for f_B as $f_B \leq \frac{c(p_l)}{\phi_B}$. The transformed conditions are summarized in Table 4.3. In this way we attain a lower bound and an upper bound of the region, in which it is optimal for retailer A to order a certain quantity at a given price. We denote the lower bound as f_{uk} and the upper bound as f_{ok}.

As obvious from Table 4.3, the dependency between optimal order quantity $q_A^*(\mathsf{s})$ and f_B, ϕ_B is different for each of the three price levels s_A. When combining the individual decisions at each price level, we need to distinguish k cases, which are different for the scenarios of no information sharing and information sharing. In each of the k cases we have a different combination of optimal order quantities at the respective price level

$$q_k^*(s) = \begin{pmatrix} q_{Ak}^*(p_l) \\ q_{Ak}^*(p_h) \\ q_{Ak}^*(r) \end{pmatrix}$$

being characterized by the domain of f_B, ϕ_B under the restriction of the critical fractile $c(\mathsf{s})$.

Table 4.3 Optimal order quantities given the opponent's promotion frequency f_B and critical fractile $c(s_A)$

Promotion price			Regular price	
$q_A^*(p_A)$	p_l	p_h	$q_A^*(r)$	r
α_A		$f_B \in \left[\frac{c(p_h)}{\phi_B}, 1\right]$	α_A	$f_B \in [c(r), 1]$
$\alpha_A + \frac{\beta_{1t}+\beta_2}{2}$	$f_B \in \left[\frac{c(p_l)}{\phi_B}, 1\right]$	$f_B \in \left[c(p_h), \frac{c(p_h)}{\phi_B}\right]$	$\alpha_A + \frac{\beta_2}{2}$	$f_B \in [0, c(r)]$
$\alpha_A + \beta_{1t} + \beta_2$	$f_B \in \left[0, \frac{c(p_l)}{\phi_B}\right]$	$f_B \in [0, c(p_h)]$		

Definition 4.1. There are k different cases for the optimal order quantity $q_{Ak}^*(s)$ as described by the critical fractile. These restrict the domain of the competitor's mixing probabilities f_B, ϕ_B to $f_B \in [f_{uk}, f_{ok}]$ and $\phi_B \in [\phi_{uk}, \phi_{ok}]$.

We shall specify the optimal order quantities and the cases k for the information scenarios respectively in Sect. 4.3.

4.2.2 Nash Equilibrium: The Optimal Promotion Frequency

So far, we have restricted our attention to the supply side, that is we have optimized the order quantity for a given price and demand function at a single retailer. Now the retailers enter the competition. We consider two symmetric retailers competing for the smart customer segment on prices. Retailers are symmetric in their loyal customer segments, i.e., $\alpha_A = \alpha_B = \alpha$ and in their promotion depth $\phi_A = \phi_B = \phi$, which is further assumed to be exogenously set.

Given a homogenous product market, the expected profits π_i of the two retailers are interdependent and the retailer who is succeeding in offering the lowest price at his store gains the smart customer segment. The other one misses out. Thus, in order to win the smart customer segment, retailers remain unpredictable to the competitor by mixing between their three price levels $s_A = \{p_l, p_h, r\}$ across stores as described by the mixed strategy profile f_B, ϕ (Narasimhan 1988).

In this environment, the decision retailer A is faced with is: what is the promotion frequency f_A that maximizes profits for a given size of the smart customer segments β_t, the optimal order quantity $q_{Ak}^*(s_A)$ as derived in stage 2 of the game and the competitor's promotion frequency f_B. Retailer A's decision must be made in the knowledge that it will affect his competitor's decision f_B which, in turn, could affect his own f_A. As common in non-corporative static games, the players simultaneously choose their promotion frequency and thereafter are committed to their chosen strategy. A concept of equilibrium for games in normal form that requires both individual rationality and mutual compatibility among the choices of the different agents was described by Nash (1951).

We take two steps in deriving the symmetric Nash equilibrium: first, we develop the expected profit under the mixed strategy f_i, ϕ. Second, we define the best response correspondence in order to characterize the symmetric Nash equilibrium for the retailer game.

Table 4.4 Profit matrix of retailer A

			Retailer B			
			$f_B\Phi$	$f_B(1-\Phi)$	$1-f_B$	
			p_l	p_h	r	
Retailer A	$f_A\Phi$	p_l	$\pi_{Ak}(p_l, p_l)$	$\pi_{Ak}(p_l, p_h)$	$\pi_{Ak}(p_l, r)$	$\pi_{Ak}(\mathrm{p}, f_B, \Phi)$
	$f_A(1-\Phi)$	p_h	$\pi_{Ak}(p_h, p_l)$	$\pi_{Ak}(p_h, p_h)$	$\pi_{Ak}(p_h, r)$	
	$1-f_A$	r	$\pi_{Ak}(r, p_l)$	$\pi_{Ak}(r, p_h)$	$\pi_{Ak}(r, r)$	$\pi_{Ak}(r, f_B, \Phi)$

4.2.2.1 Expected Profit in Mixed Strategies

In this game, the retailer A's profit depends on (1) his optimal order quantity as given by the newsvendor solution $q_{Ak}(\mathrm{s}_A)^*$, (2) his own strategy f_A and (3) the opponent's strategy f_B. Table 4.4 displays the pure strategy profits of retailer A for all possible pure strategy choices at a given optimal order quantity $q^*_{Ak}(\mathrm{s}_A)$ $\forall f_B \in [f_{uk}, f_{ok}]$ and $\phi \in [\phi_{uk}, \phi_{ok}]$, including the interdependency of own profits from the opponent's strategy. For example, retailer A's profit when he plays the low promotion price p_l and retailer B plays his regular price r is $\pi_{Ak}(p_l, r)$, where k marks the respective case as characterized in Definition 4.1. Extending (4.2) by the interdependencies, retailer A attains an expected pure strategy profit of

$$\pi_{Ak}(\mathrm{s}_A, \mathrm{s}_B) = (\mathrm{s}_A - w)d_A(\mathrm{s}_A, \mathrm{s}_B) - c_u(\mathrm{s}_A)\max\left[d_A(\mathrm{s}_A, \mathrm{s}_B) - q^*_{Ak}(\mathrm{s}_A), 0\right] \\ - c_o \max\left[q^*_{Ak}(\mathrm{s}_A) - d(\mathrm{s}_A, \mathrm{s}_B), 0\right].$$

The expected profit for an arbitrary mixed strategy profile f_A, f_B, ϕ is found by weighting each of A's pure strategy profits by the probability of its occurrence, as determined by the mixed strategy profile f_A, f_B, ϕ ($\forall f_B \in [f_{uk}, f_{ok}]$ and $\phi \in [\phi_{uk}, \phi_{ok}]$). The expected profit is then the weighted sum over all possible pure strategy profiles s_A. Again, we assume the promotion depth ϕ to be exogenously set and equal for both retailers.

Then, we can write the expected profit of retailer A in case k as

$$\pi_{Ak}(f_A; f_B, \phi) = f_A \pi_{Ak}(\mathrm{p}; f_B, \phi) + (1 - f_A)\pi_{Ak}(r; f_B, \phi). \tag{4.4}$$

The semicolon in the above equation is used to denote that, while f_A is the decision variable for retailer A, the other parameters are outside of his control. In the following section, we will first consider the expected profit in promotions $\pi_{Ak}(\mathrm{p}; f_B, \phi)$ and then focus on the expected profit at the regular price $\pi_{Ak}(r; f_B, \phi)$. Both of these are indicated in Table 4.4 for clarification.

The expected profit at the promotion price $\pi_{Ak}(p; f_B, \phi)$ can be formulated as

$$\begin{aligned}\pi_{Ak}(\mathsf{p}; f_B, \phi) = &\ \phi\pi(p_l, r) + (1-\phi)\pi(p_h, r)\\ &- f_B\Big(\phi\big(\pi(p_l, r) - \phi\pi(p_l, p_l) - (1-\phi)\pi(p_l, p_h)\big)\\ &+ (1-\phi)\big(\pi(p_h, r) - \phi\pi(p_h, p_l) - (1-\phi)\pi(p_h, p_h)\big)\Big).\end{aligned}$$

In order to simplify notation, we introduce the following vectors, describing the probability, and the underage cost

$$\sigma = \begin{pmatrix}\phi\\ 1-\phi\end{pmatrix},\ \chi = \begin{pmatrix}\phi\\ 1+\phi\end{pmatrix},\ \mathsf{p} = \begin{pmatrix}p_l\\ p_h\end{pmatrix},\ c_u(\mathsf{p}) = \begin{pmatrix}c_u(p_l)\\ c_u(p_h)\end{pmatrix}$$

and can rewrite previous equation as

$$\pi_{Ak}(\mathsf{p}; f_B, \phi) = \sigma\pi(\mathsf{p}, r) - f_B\sigma\big(\pi_{Ak}(\mathsf{p}, r) - \chi\pi_{Ak}(\mathsf{p}, \mathsf{p})\big).$$

Further, we define a vector describing the optimal order quantities of promotions as

$$q^*_{Ak}(\mathsf{p}) = \begin{pmatrix}q^*_{Ak}(p_l)\\ q^*_{Ak}(p_h)\end{pmatrix}$$

and derive the inventory vector on promotions as

$$\begin{aligned}\Gamma_{Ak}(\mathsf{p}) = &\ c_u(\mathsf{p})\Big((1-\phi)\max[d(\mathsf{p}, p_h) - q^*_{Ak}(\mathsf{p}), 0] - \max[d(\mathsf{p}, r) - q^*_{Ak}(\mathsf{p}), 0]\Big)\\ &+ c_o\Big(\phi\max[q^*_{Ak}(\mathsf{p}) - d(\mathsf{p}, p_l), 0] + (1-\phi)\max[q^*_{Ak}(\mathsf{p}) - d(\mathsf{p}, p_h), 0]\Big)\end{aligned} \tag{4.5}$$

while considering that underage cost $c_u(\mathsf{p})$ can only occur if the opponent offers at p_h or r, whereas overage cost c_o can only occur if the opponent offers at p_l or p_h.

Then, we can rewrite

$$\pi_{Ak}(\mathsf{p}; f_B, \phi) = \Upsilon(\mathsf{p}) + \Lambda_k(\mathsf{p}) - f_B\Omega_k(\mathsf{p}), \tag{4.6}$$

where

$$\Upsilon(\mathsf{p}) = \sigma\alpha(\mathsf{p} - w), \tag{4.7}$$

$$\Lambda_k(\mathsf{p}) = \sigma\big(\beta_{1t} + \beta_2\big)(\mathsf{p} - w) - \sigma c_u(\mathsf{p})\max[d(\mathsf{p}, r) - q^*_{Ak}(\mathsf{p}), 0], \tag{4.8}$$

$$\Omega_k(\mathsf{p}) = \sigma\chi\frac{\beta_{1t} + \beta_2}{2}(\mathsf{p} - w) + \sigma\Gamma_k(\mathsf{p}). \tag{4.9}$$

The expected profit at the promotion price $\pi_{Ak}(\mathsf{p}; f_B, \phi)$ is a composition of the BASE profit $\Upsilon(\mathsf{p})$ plus the CHANCE from offering a promotion price $\Lambda_k(\mathsf{p})$ less the competition RISK $\Omega_k(\mathsf{p})$ if both the retailer and his competitor are promoting.

The BASE profit of (4.7) represents the riskless profit from the loyal customer segment when offering a promotion. Note that neither underage nor overage cost incur as the retailer's order quantity amounts to the size of his loyal customer segment α at the minimum and any increase in the order quantity would aim at serving the smart customer segment.

Equation (4.8) defines the CHANCE of a promotion $\Lambda_k(p)$, as the profit a retailer attains in case the opponent offers at regular price: the retailer gains the complete smart customer segments and attains sales of $(\beta_{1t} + \beta_2)\sigma(\mathsf{p} - w)$. However, if the retailer does not provide the full service level to the smart customer segments, he incurs underage cost amounting to $\sigma c_u(\mathsf{p}) \max[d(\mathsf{p}, r) - q^*_{Ak}(\mathsf{p}), 0]$. The promotion CHANCE becomes negative in case of a low service level, when underage cost exceeds the margin of incremental sales.

The profit impact of competitor's promotions is described as the RISK of a promotion $\Omega_k(\mathsf{p})$ and captured in (4.9). If both the retailer and his competitor offer a promotion price, the retailer might lose sales to the smart customer segments adding up to $\sigma\chi\frac{\beta_{1t}+\beta_2}{2}(\mathsf{p} - w)$. As evident from $\Gamma_k(\mathsf{p})$, if the retailer provided a high service level, he additionally has high overage cost at the end of the period. On the contrary, if he provides a low service level, he benefits from the competitor's promotion: in this case, he has lower stockouts the more frequently the competitor promotes and the promotion RISK is negative.

Similar to the profit at the promotion price, we decompose the profit at the regular price $\pi_{Ak}(r; f_B, \phi)$:

$$\pi_{Ak}(r; f_B, \phi) = \pi_{Ak}(r, r) - f_B\big(\pi_{Ak}(r, r) - \phi\pi_{Ak}(r, p_l) - (1 - \phi)\pi_{Ak}(r, p_h)\big),$$

which we can rewrite as

$$\pi_{Ak}(r; f_B) = \Upsilon(r) + \Lambda_k(r) - f_B\Omega_k(r), \tag{4.10}$$

where

$$\Upsilon(r) = \alpha(r - w), \tag{4.11}$$

$$\Lambda_k(r) = \frac{\beta_2}{2}(r - w) - \sigma c_u(r) \max[d(r, r) - q^*_{Ak}(r), 0], \tag{4.12}$$

$$\Omega_k(r) = \frac{\beta_2}{2}(r - w) + \Gamma_k(r), \tag{4.13}$$

with the inventory vector at the regular price $\Gamma_k(r)$ being derived as

$$\Gamma_k(r) = -c_u(r) \max[d(r, r) - q^*_{Ak}(r), 0] + c_o \max[q^*_{Ak}(r) - d(r, r), 0]. \tag{4.14}$$

Also the profit at the regular price can be expressed as the sum of the BASE profit $\Upsilon(r)$ plus the CHANCE from offering at the regular price $\Lambda_k(r)$ less the competition RISK $\Omega_k(r)$ if the competitor is offering at the promotion price.

In this case, the BASE profit amounts to $\alpha(r-w)$ as given by (4.11). The CHANCE at the regular price is characterized by (4.12) and defines the situation where both the retailer and his competitor offer at the regular price. The smart customer segment is restricted to the store-switching customer segment, whose demand is split equally amongst the retailers, raising profits by $\frac{\beta_2}{2}(r-w)$. The retailer risks underage cost of $c_u(r)\max[d(r,r) - q^*_{Ak}(r), 0]$ for not providing the full service level.

Equation (4.13) represents the RISK from offering a regular price: promotions by the competitor result in the loss of the store-switching customers, and in lower sales of $\frac{\beta_2}{2}(r-w)$. As evident from the inventory vector at the regular price $\Gamma_k(r)$, in case of a full service level, the retailer faces excess stock at the end of the period and pays overage cost. However, in case of a low service level, the retailer benefits from competition: a successful promotion by the competitor therefore reduces the stockout risk and hence reduces the retailer's underage cost.

4.2.2.2 Best Response

Each retailer will choose the promotion frequency which maximizes expected profits given his beliefs about what actions the other retailer will take. That is, each retailer plays his best response, denoted as $\delta_{Ak}(f_B)$, to counter his opponent. We will show in detail how to compute retailer A's best response. Retailer B's is derived in exactly the same way.

Definition 4.2. The best response of retailer A $\delta_{Ak}(f_B)$ is the promotion frequency that maximizes his expected profit $\pi_{Ak}(f_A; f_B, \phi)$ while taking the other retailer's strategy $f_B \in [f_{uk}, f_{ok}]$ as given. Formally, we write

$$\delta_{Ak}(f_B) = \max_{f_A \in [0,1]} \pi_{Ak}(f_A; f_B, \phi), \quad \forall f_B \in [f_{uk}, f_{ok}].$$

The first order condition is calculated from (4.4) as

$$\begin{aligned}\delta_{Ak}(f_B) &= \frac{\partial \pi_{Ak}(f_A; f_B, \phi)}{\partial f_A} = \pi_{Ak}(p; f_B, \phi) - \pi_{Ak}(r; f_B, \phi)\\ &= -\Delta\Upsilon + \Delta\Lambda_k - f_B \Delta\Omega_k,\end{aligned} \tag{4.15}$$

where

$$\begin{aligned}\Delta\Upsilon &= \Upsilon(r) - \Upsilon(p),\\ \Delta\Lambda_k &= \Lambda_k(p) - \Lambda_k(r),\\ \Delta\Omega_k &= \Omega_k(p) - \Omega_k(r).\end{aligned}$$

Observe that the difference of BASE profits from the regular and promotion price is defined negatively, given that in promotions, the retailer sells the product at a lower than regular price to his loyal customers. Further, the retailer's best response trades off the CHANCE of a promotion price versus a regular price and the associated RISK respectively.

Evidently, the best response is a balance between the BASE profit and the CHANCE from offering at promotion or at regular price on the one hand, and the RISK from offering at promotion or at regular price on the other hand.

This balance is influenced by two important factors: (1) the size of the stockpiling segment β_{1t} relative to the size of the store-switching β_2 and loyal customer segment α and (2) the competitor's promotion frequency f_B relative to the retailer's cost ratio, expressed by the critical fractile $c(\mathsf{s})$.

The incentive for the retailer to offer a promotion price is to win the stockpiling segment. Whenever this segment exceeds a critical size, denoted by τ, the retailer's best response is positive and it is profitable for the retailer to promote. In contrast, whenever the segment's size falls below the critical size, it is optimal for retailer A to protect his profits from the loyal customer segment α and speculate on attaining half of the store-switching customer segment β_2 at the regular price. We use the following definition to describe the critical size τ for the stockpiling segment:

Definition 4.3. There exists a critical size for the stockpiling segment, denoted by $\tau > 0$ at which the retailer's best response changes from playing the regular price r to playing the promotion price p.

We shall provide the precise values for τ in the scenario no information sharing and information sharing in the following chapter.

Further, the best response depends on the competitor's promotion frequency f_B, which determines both the optimal order quantity (as described in the previous chapter) and the best response of retailer A: whenever the competitor promotes frequently, it is optimal for retailer A to reduce his order quantity. This however increases his dependency on retailer B – he will only make positive profits in case the opponent remains with the high promotion frequency, otherwise he faces cost from stockouts.

In order to capture the importance of both factors on the best response, we include β_{1t} as

$$\delta_{Ak}(f_B, \beta_{1t}).$$

Combining the influential factors with the BASE, CHANCE and RISK terminology, we can determine three general best responses for the retailer. Firstly, if the size of the stockpiling segment β_{1t} is too small, a situation can occur in which the RISK is larger than the BASE and CHANCE elements of the best response and it becomes optimal for retailer A to offer at regular price across all stores with $f_A = 0$, independent of the competitor's promotion frequency f_B.

Secondly, if the size of the stockpiling segment β_{1t} exceeds the critical value τ, retailer A's best response becomes dependent on the competitors's promotion frequency: he plays strategically. For every f_B that allows the RISK to exceed the BASE and CHANCE elements of the best response ($\delta_A(f_B, \beta_{1t}) < 0$), retailer A stays out of the promotion competition and offers at the regular price r. On the contrary, if f_B reduces the RISK and increases the BASE and CHANCE ($\delta_A(f_B, \beta_{1t}) > 0$), retailer A's best response is to offer at the promotion price p. In between these two pure strategies, the competitor's promotion frequency f_B exactly balances BASE and CHANCE on the one hand, and RISK on the other hand ($\delta_A(f_B, \beta_{1t}) = 0$). In this case, retailer A's best response is to mix between his pure strategies.

Thirdly, if the size of the stockpiling segment β_{1t} becomes even larger, it becomes more profitable for retailer A to enter promotion competition in order to gain a piece of the stockpiling pie. In this third case, playing the pure strategy "promotion" is better than "regular", independent of the competitor's promotion frequency f_B. We neither find this case in the scenario of no information sharing, nor in the scenario of information sharing. Consequently, we will not further consider this case.

We can formalize the above arguments in the following proposition.

Proposition 4.1. *The best response of retailer A $\delta_{Ak}(f_B, \beta_{1t})$ changes with the size of the stockpiling segment β_{1t} and the competitor's promotion frequency f_B. We differentiate two cases:*

Case i: *If $\beta_{1t} < \tau$, then $\delta_{Ak}(f_B, \beta_{1t}) < 0 \; \forall f_B \in [f_{uk}, f_{ok}]$ and retailer A's best response is to play his pure strategy "regular" no matter how retailer B plays: He has a dominant strategy.*

$$f_A^* = 0.$$

Case ii: *If $\beta_{1t} > \tau$ retailer A's best response is to play strategically and mix between his pure strategies "regular" and "promotion" with probability*

$$f_A^* = \begin{cases} 0 & \text{if } \delta_{Ak}(f_B, \beta_{1t}) < 0, \\ [0, 1] & \text{if } \delta_{Ak}(f_B, \beta_{1t}) = 0 \Rightarrow f_{Bk}^{\dagger}, \\ 1 & \text{if } \delta_{Ak}(f_B, \beta_{1t}) > 0, \end{cases}$$

with

$$f_{Bk}^{\dagger} = \frac{-\Delta\Upsilon + \Delta\Lambda_k}{\Delta\Omega_k}.$$

As evident from the proposition, retailer A is only playing a mixed strategy in case (1) the stockpiling segment is large and thus profitable enough and (2) his competitor plays at a frequency $f_{Bk}^{\dagger}$ that exactly balances his BASE and CHANCE profit with the RISK profit. In all other cases, he sticks to playing one of his two pure strategies "promotion" or "regular".

In order to identify whether it is optimal for the retailer to play (1) the promotion price, (2) the regular price or (3) to mix between the two prices, we use a three step approach for each case k. The procedure is visualized in Fig. 4.4 and described in the following:

1. $\delta_{Ak}(f_B = 0)$ We begin by considering the best response of retailer A in case k, if the opponent does not enter promotion competition, but instead offers at the

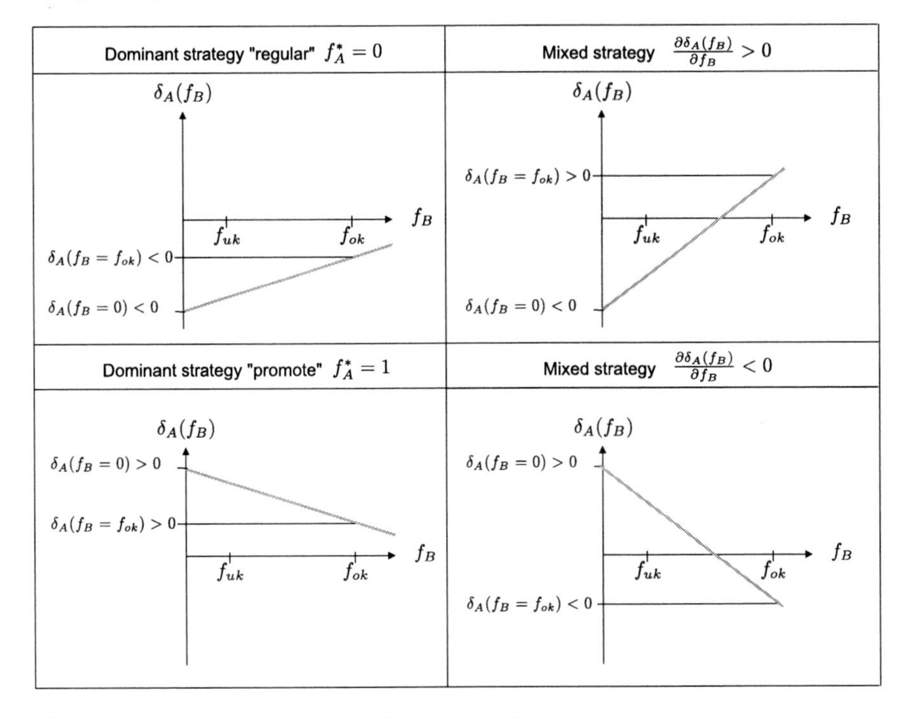

Fig. 4.4 Best response for dominant and mixed strategies

regular price, i.e., $f_B = 0$. We analyze, whether for

$$\delta_{Ak}(f_B = 0) = -\Delta\Upsilon + \Delta\Lambda_k$$

the BASE and CHANCE are positive or negative.

2. $\delta_{Ak}(f_B = f_{ok})$ In a next step, we analyze the best response of retailer A in case k if the opponent plays the upper bound of the domain, i.e., $f_B = f_{ok}$. Again, we analyze whether

$$\delta_{Ak}(f_B = f_{ok}) = -\Delta\Upsilon + \Delta\Lambda_k - f_{ok}\Delta\Omega_k \tag{4.16}$$

is positive or negative.

3. Comparison of (1) and (2): if the best response at both the upper bound and the zero bound have the same sign, we can conclude that the retailer has a dominant strategy. It is dominant "regular" if both the best response at $f_B = 0$ and at $f_B = f_{ok}$ is negative and dominant "promote" if both values are positive. To the contrary, if the sign changes, a frequency f_B exists at which $\delta_{Ak}(f_B, \beta_{1t}) = 0$ and it is optimal for the retailer to play strategically and to mix between his two pure strategies, promotion and regular.

Finally, we can draw the best response of retailer A in a reaction correspondence diagram: it shows the best response of retailer A for any promotion frequency of retailer B, $f_B \in [f_{uk}, f_{ok}]$. We shall provide these figures in the respective sections of the scenarios no information sharing and information sharing (compare Chaps. 4.3 and 4.4).

4.2.2.3 Symmetric Nash Equilibrium

A Nash equilibrium is a profile of strategies such that each retailer's strategy is a best response to the other retailer's strategies. Consequently, no other combination of strategies allows the retailer to earn a higher profit.

Definition 4.4. A mixed strategy profile f_A^*, f_B^* is a Nash equilibrium, if for all players

$$\pi(f_A^*, f_B^*, \phi) > \pi(f_A, f_B^*, \phi).$$

A pure strategy Nash equilibrium is a pure strategy profile that satisfies the same conditions.

Observe that a pure strategy Nash equilibrium can be interpreted as an equilibrium in degenerated mixed strategies, in the sense that the retailer puts all his weight on one pure strategy. In contrast, in a mixed strategy equilibrium, the retailer puts positive weight on more than one pure strategy. Further, in a mixed strategy equilibrium, a retailer chooses his strategy in a way that makes his competitor indifferent between any of his two pure strategies "promotion" and "regular" (Fudenberg and Tirole 1998).

Our objective is to identify the Nash equilibrium for symmetric retailers. The previous section revealed that the best response of the retailers can either be a pure strategy or a mixed strategy, depending on the size of the stockpiling segment β_{1t} relative to the other segments β_2, α. This allows us to make the following proposition.

Proposition 4.2 (Nash equilibrium). *The game has a symmetric Nash equilibrium $f^* = f_A^* = f_B^*$, if each retailer plays his best response as described by Proposition 4.1. The equilibrium depends on the size of the stockpiling segment β_{1t}. We differentiate two cases:*

Case i: *If $\beta_{1t} < \tau$, the game has a symmetric pure strategy equilibrium at $f^* = 0$. Consequently both retailers have a dominant strategy "regular".*

Case ii: *If $\beta_{1t} > \tau$ the game has a symmetric mixed strategy equilibrium at*

$$f^* = \frac{-\Delta\Upsilon + \Delta\Lambda_k}{\Delta\Omega_k}.$$

Proof. Follows directly from Definition 4.3 and Proposition 4.1. □

Intuitively, the equilibrium promotion frequency f^*, for a significant size of the stockpiling customer segment ($\beta_{1t} > \tau$), depends on the ratio of BASE and CHANCE versus the RISK of a promotion. It is optimal for the retailer to promote more frequently, if (1) the promotion RISK is reduced and (2) the BASE profit and the CHANCE of a promotion increase.

We ensure the existence of the Nash Equilibrium in the following proposition.

Proposition 4.3. *A mixed strategy Nash equilibrium exists if the pure strategy sets are nonempty compact subsets of Euclidean space and payoffs are continuous.*

Proof. Note that no quasiconcavity requirement on payoffs and no convexity requirement on strategy spaces are necessary due to the convexifying effect of mixed strategies on best responses. The payoffs and strategy spaces of the mixed extensions have the appropriate convexity properties.

The proof proceeds by showing that the strategy sets of the mixed extensions f_i are convex and compact, which is fulfilled in our case. Furthermore, the expected profit of retailer A $\pi_A(f_A; f_B, \phi)$ is linear in f_A and therefore quasiconcave in f_A and continuous in f_A, f_B (Vives 1999 p. 44). Existence follows then from the standard fixed point theorem (Fudenberg and Tirole 1998 p. 29). □

Finally, we provide the retailer's profit in equilibrium from (4.4) as

$$\beta_{1t} < \tau \quad \pi_{Ak}(0,0,\phi) = \left(\alpha + \frac{\beta_2}{2}\right)(r - w), \tag{4.17}$$

$$\begin{aligned} \beta_{1t} > \tau \quad \pi_{Ak}(f_A^*, f_B^*, \phi) &= f_A^* \pi_{Ak}(p) + (1 - f_A^*)\pi_{Ak}(r) \\ &= \Upsilon(r) + \Lambda_k(r) - f_B^* \Omega_k(r). \end{aligned} \tag{4.18}$$

4.3 Information Scenarios

The objective is to identify the impact of sharing the Competition Index. Therefore, we shall apply the previously described setup and the retailer's optimization problems to two different scenarios in this section.

The first scenario describes the base case, where no information is shared. It is named "no information sharing" scenario and the respective variables are denoted with $(\hat{\cdot})$. In the second case, the retailer receives information from the Competition Index before he places his order. This second scenario is named "information sharing" scenario and the respective variables are denoted with $(\check{\cdot})$. Before describing the two scenarios, we will provide insights into how the retailer can exploit the information from the Competition Index.

4.3.1 Information from the Competition Index

With the information from the Competition Index, each individual retailer in a two retailer game, can conclude on whether his competitor offers the product at regular or at promotion price.

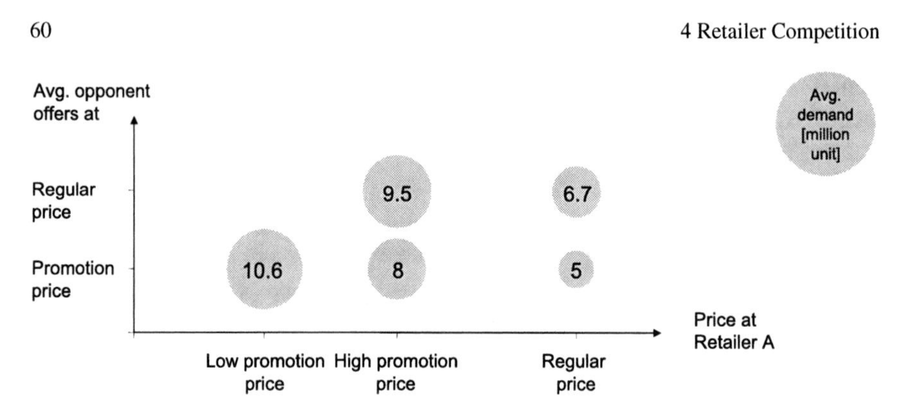

Fig. 4.5 Average weekly demand given the price of an average opponent. Source: own data

The information revealed from the Competition Index, i.e., whether the competitor offers the product at promotion or at regular price, is represented on the y-axis of Fig. 4.5. For each possible price level of the retailer himself (low promotion price, high promotion price and regular price), on the x-axis, the size of the bubble reveals the average demand for the product, given the opponents pricing.

For example, in case the retailer decided to offer the product at a high promotion price, he would face an average demand of 9.5 million units in case the opponent asks for the regular price. The average demand decreases to 8 million units provided the opponent offers the product at promotion price.

On attaining the information, we can see two different effects in the retailer's reaction, characterized as price and supply game. Whether the retailer plays the price or the supply game depends on the timing of information sharing. Both games are evident from Fig. 4.5.

The first effect of the Competition Index is an improved inventory management: due to the information, the demand uncertainty is reduced and orders are adapted more accurately to actual demand, resulting in both lower out-of-stocks and less overstocking. We denote this first effect as supply game.

A second effect of attaining the information, that the competitor is offering at promotion price, supports the retailer's desire to maximize his profit. By further decreasing his promotion price from a high to a low level, the average demand increases to 10.6 million units. The second effect is characterized as price game.

Whether the retailers play either of the two games (price or supply) or both games simultaneously, depends on the timing of information sharing as described earlier in Fig. 4.1. Here it is laid down that retailers first place their promotion frequency decision and only thereafter their order decision. In case the retailer receives the Competition Index after setting his promotion prices but before ordering, he is restricted to playing the supply game. However, if he receives the Competition Index before setting promotion prices, he can adapt both: pricing and order quantity.

We shall assume that manufacturers provide the information regarding the competitive pressure only after the retailer's promotion frequency decision, and manifest this assumption with two arguments. Firstly, the closer to the time of the promotions the manufacturer receives the information from the retailer, the more detailed and the more accurate is the information. Hence, the manufacturer's calculation of the Competition Index has a more solid base making the information from the Competition Index more meaningful for the retailer.

Secondly, the manufacturer does not support suspicious price wars amongst competitors by allowing the retailers to enter the price game. Instead, if the Competition Index is shared before the retailer places the order quantity, the information can only be employed to streamline supply chain efficiency: with this information, a retailer is aware whether the opponent is offering the product at regular or promotion price. Reducing the pricing uncertainty at the same time reduces the demand uncertainty and hence orders are more adapted to actual demand. This in turn improves supply chain efficiency.

4.3.2 No Information Sharing Scenario

In the scenario of no information sharing, the retailer does not attain any information on the competitor's pricing strategy. This scenario serves as the base case to allow comparison with the information sharing scenario which is described in the next chapter. We begin this section by describing the retailer's demand, his optimal order quantity and profit under a mixed strategy. From this, we derive the retailer's best response in order to determine the symmetric mixed strategy equilibrium.

Cumulative Distribution Function of Demand

From the demand, as defined in Table 4.1, the cumulative distribution function of demand $\hat{\Theta}(d(\mathsf{s}_A))$ for a given price of the retailer's strategy space $\mathsf{s}_A = \{p_l, p_h, r\}$ is summarized in Table 4.5 and drawn in Fig. 4.6. The cumulative distribution function describes the probability that demand does not exceed a certain level given the uncertainty regarding the competitor's promotion frequency $\hat{f}_B$ and depth ϕ.

For example, in case retailer A offers the product at the regular price, the demand α will occur with probability f_B. The demand can reach a maximum of $\alpha + \frac{\beta_2}{2}$ with a cumulative distribution of 1. Figure 4.6 shows how the distribution function increases in discrete steps, reflecting the three possible discrete demand levels α, $\alpha + \frac{\beta_{1t}+\beta_2}{2}$ and $\alpha + \beta_{1t} + \beta_2$ in the case of promotion and two discrete demand levels α and $\alpha + \frac{\beta_2}{2}$ in the case of regular price.

Table 4.5 Optimal order quantities and cumulative densities for retailer A – no information sharing

Promotion price			Regular price	
$q_A^*(\mathsf{p}_A)$	$\Theta_A(d(p_l))$	$\Theta_A(d(p_h))$	$q_A^*(r)$	$\Theta_A(d(r))$
α	0	$f_B\phi_B$	α	f_B
$\alpha + \frac{\beta_{1t}+\beta_2}{2}$	$f_B\phi_B$	f_B	$\alpha + \frac{\beta_2}{2}$	1
$\alpha + \beta_{1t} + \beta_2$	1	1		

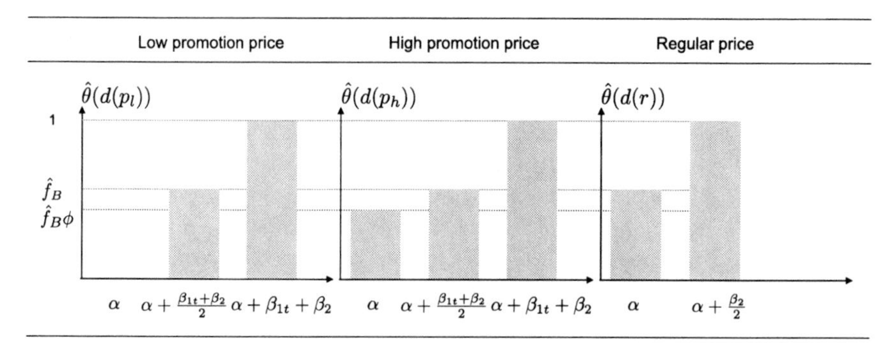

Fig. 4.6 Cumulative distribution function of demand – no information sharing

Optimal Order Quantity

The optimal order quantity of a retailer is determined by balancing the risk of overstocking and understocking as described by the critical fractile $c(\mathsf{s}_A)$ in (4.3). For a given price level of the retailer's strategy space s_A, we obtain the retailer's optimal order quantity $\hat{q}_A^*(\mathsf{s}_A)$ by determining the value of $\hat{q}_A(\mathsf{s}_A)$ that makes the cumulative distribution function of demand $\hat{\Theta}\big(d(s_A)\big)$ equal to the critical fractile $c(\mathsf{s}_A)$. Given Table 4.5, we attain the following cases for each price level for a given promotion frequency of retailer B:

$$\hat{q}_A^*(p_l) = \begin{cases} \alpha + \frac{\beta_{1t}+\beta_2}{2} & \hat{f}_B \in \left[\frac{c(p_l)}{\phi}, 1\right], \\ \alpha + \beta_{1t} + \beta_2 & \hat{f}_B \in \left[0, \frac{c(p_l)}{\phi}\right). \end{cases} \tag{4.19}$$

$$\hat{q}^*(p_h) = \begin{cases} \alpha & \hat{f}_B \in \left[\frac{c(p_h)}{\phi}, 1\right], \\ \alpha + \frac{\beta_{1t}+\beta_2}{2} & \hat{f}_B \in \left[c(p_h), \frac{c(p_h)}{\phi}\right), \\ \alpha + \beta_{1t} + \beta_2 & \hat{f}_B \in [0, c(p_h)), \end{cases} \tag{4.20}$$

$$\hat{q}^*(r) = \begin{cases} \alpha & \hat{f}_B \in \left[c(r), 1\right], \\ \alpha + \frac{\beta_2}{2} & \hat{f}_B \in \left[0, c(r)\right). \end{cases} \tag{4.21}$$

Next, we combine the optimal order quantities at the respective price level while taking into consideration the domain of the critical fractile.

Applying Definition 4.1, we combine the individual domains. This results in $k = 8$ different combinations of optimal order quantities based on the domains of the critical fractile $c(\mathsf{s}_A)$. Table 4.6 summarizes the domain for $\hat{f}_B, \phi$ and the order quantities for the respective cases. Observe that we focus on the minimum and the maximum value of the domain of $\hat{f}_B, \phi$ respectively in order to maintain clarity.

Figure 4.7 visualizes the different domains. The x-axis shows the exogenously set promotion depth ϕ with the bounds $0, c(p_l), c(p_h), \frac{c(p_l)}{c(r)}, \frac{c(p_h)}{c(r)}, \frac{c(p_l)}{c(p_h)}$ and 1 as

Table 4.6 Domains for the promotion frequency $\hat{f}_B$ and depth ϕ and the respective optimal order quantities – no information sharing

Case k	$\hat{f}_B \in$	$\phi \in$	$\hat{q}^*(p_l)$	$\hat{q}^*(p_h)$	$\hat{q}^*(r)$
1	$\left[\frac{c(p_h)}{\phi}, 1\right]$	$[c(p_h), 1]$	$\alpha + \frac{\beta_{1t}+\beta_2}{2}$	α	α
2	$\left[\frac{c(p_h)}{\phi}, c(r)\right]$	$\left[\frac{c(p_h)}{c(r)}, 1\right]$	$\alpha + \frac{\beta_{1t}+\beta_2}{2}$	α	$\alpha + \frac{\beta_2}{2}$
3	$\left[\frac{c(p_l)}{\phi}, c(p_h)\right]$	$\left[\frac{c(p_l)}{c(p_h)}, 1\right]$	$\alpha + \frac{\beta_{1t}+\beta_2}{2}$	$\alpha + \beta_{1t} + \beta_2$	$\alpha + \frac{\beta_2}{2}$
4	$[c(p_h), c(r)]$	$\left[\frac{c(p_l)}{c(r)}, 1\right]$	$\alpha + \frac{\beta_{1t}+\beta_2}{2}$	$\alpha + \frac{\beta_{1t}+\beta_2}{2}$	$\alpha + \frac{\beta_2}{2}$
5	$[c(r), 1]$	$\left[c(p_l), \frac{c(p_h)}{c(r)}\right]$	$\alpha + \frac{\beta_{1t}+\beta_2}{2}$	$\alpha + \frac{\beta_{1t}+\beta_2}{2}$	α
6	$[c(r), 1]$	$\left[0, \frac{c(p_l)}{c(r)}\right]$	$\alpha + \beta_{1t} + \beta_2$	$\alpha + \frac{\beta_{1t}+\beta_2}{2}$	α
7	$[c(p_h), c(r)]$	$\left[0, \frac{c(p_l)}{c(p_h)}\right]$	$\alpha + \beta_{1t} + \beta_2$	$\alpha + \frac{\beta_{1t}+\beta_2}{2}$	$\alpha + \frac{\beta_2}{2}$
8	$[0, c(p_h)]$	$[0, 1]$	$\alpha + \beta_{1t} + \beta_2$	$\alpha + \beta_{1t} + \beta_2$	$\alpha + \frac{\beta_2}{2}$

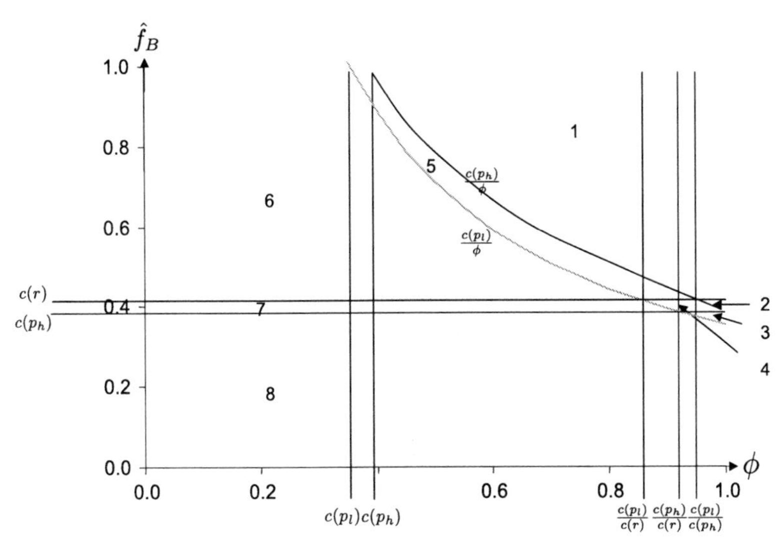

Fig. 4.7 Domains of cases for the optimal order quantity – no information sharing

given in Table 4.6. On the y-axis the competitor's promotion frequency $\hat{f}_B$ is plotted along with its bounds $0, c(p_h), c(r), \frac{c(p_h)}{\phi}, \frac{c(p_l)}{\phi}$ and 1. The resulting domains for the $k = 8$ cases are marked in the Fig. 4.7.

The order quantity decreases in the competitor's promotion frequency $\hat{f}_B$ and depth ϕ, relative to the critical fractile: the more frequently the opponent promotes, the lower the probability of attaining the switching segments and consequently the lower the service level, i.e., the lower the order quantity.

In Case 1, the optimal order quantity is the smallest across all price levels p_l, p_h and r, due to the low probability of winning the smart customer segments given the high promotion frequency $\hat{f}_B$ and depth ϕ of retailer B relative to the respective critical fractile. If retailer B reduces his promotion frequency $\hat{f}_B$ to below $c(r)$, it is optimal for retailer A to speculate for a tie at the regular price and increase his order quantity at the regular price. Also, if the promotion frequency of retailer B falls below $\frac{c(p_l)}{\phi}$, the probability of winning the smart customer segments at the low promotion price p_l increases, and consequently, it is optimal for retailer A to increase the order quantity at the respective price level. If he offers the product at the high promotion price p_h, retailer A's chance of winning the smart customer segments increases if the competitor's promotion frequency drops below $\frac{c(p_h)}{\phi}$ and $c(p_h)$ respectively. As a result, retailer A will in turn increase his order quantities. Finally, in Case 8, the optimal order quantity of retailer A is highest, providing full service level to all three customer segments α, β_{1t} and β_2 across all three price levels p_l, p_h and r, based on the highest possible demand.

Expected Profit Under Mixed Strategies

The optimal order quantity defines the domain of the mixing probabilities $\hat{f}_B, \phi$ as shown in Table 4.6. Within these domains, we determine the expected profit under mixed strategies from (4.4), (4.6) and (4.10) for each of the $k = 8$ cases as

$$\hat{\pi}_{Ak}(\hat{f}_A; \hat{f}_B, \phi) = \hat{f}_A \hat{\pi}_{Ak}(\mathsf{p}; \hat{f}_B, \phi) + (1 - \hat{f}_A)\hat{\pi}_{Ak}(r; \hat{f}_B, \phi), \tag{4.22}$$

where

$$\hat{\pi}_{Ak}(\mathsf{p}; \hat{f}_B, \phi) = \Upsilon(\mathsf{p}) + \hat{\Lambda}_k(\mathsf{p}) - \hat{f}_B \hat{\Omega}_k(\mathsf{p}) \tag{4.23}$$

and

$$\pi_{Ak}(r; \hat{f}_B) = \Upsilon(r) + \hat{\Lambda}_k(r) - \hat{f}_B \hat{\Omega}_k(r) \tag{4.24}$$

with the BASE profits $\Upsilon(\mathsf{p}), \Upsilon(r)$, the CHANCES $\hat{\Lambda}_k(\mathsf{p}), \hat{\Lambda}_k(r)$ and the RISKS $\hat{\Omega}_k(\mathsf{p}), \hat{\Omega}_k(r)$ as defined in (4.7)–(4.9) and (4.11)–(4.13).

The profit and its components BASE, CHANCE and RISK for the $k = 8$ cases are summarized in Table 4.7.

Best Response

Retailer A's best response is the promotion frequency $\hat{f}_A$ which produces the most favorable immediate outcome, taking the other retailer's promotion frequency $\hat{f}_B$ as given.

From (4.15), we can derive the best response on no information sharing as

$$\hat{\delta}_{Ak}(\check{f}_B) = -\Delta\Upsilon + \Delta\hat{\Lambda} - \hat{f}_B \Delta\hat{\Omega}_k. \tag{4.25}$$

Table 4.7 Overview of profit components – no information sharing

Case k	BASE	CHANCE	RISK
	$\alpha \Upsilon(p)$	$+(\beta_{1t}+\beta_2)\hat{\Lambda}_k(p)$	$-\hat{f}_B(\beta_{1t}+\beta_2)\hat{\Omega}_k(p)$
1, 2	$\phi p_l + (1-\phi)p_h - w$	$\phi p_l + (1-\phi)p_h - w - \phi\frac{c_u(p_l)}{2} - (1-\phi)c_u(p_h)$	$\frac{1}{2}\big(\phi^2 p_l + (1-\phi^2)p_h - w - \phi^2 c_u(p_l) - (1-\phi^2)c_u(p_h)\big)$
3	$\phi p_l + (1-\phi)p_h - w$	$\phi p_l + (1-\phi)p_h - w - \phi\frac{c_u(p_l)}{2}$	$\frac{1}{2}\big(\phi^2 p_l + (1-\phi^2)p_h - w - \phi^2 c_u(p_l) + (1-\phi^2)c_o\big)$
4, 5	$\phi p_l + (1-\phi)p_h - w$	$\phi p_l + (1-\phi)p_h - w - \phi\frac{c_u(p_l)}{2} - (1-\phi)\frac{c_u(p_h)}{2}$	$\frac{1}{2}\big(\phi^2 p_l + (1-\phi^2)p_h - w - \phi^2 c_u(p_l) - (1-\phi)(c_u(p_h) - \phi c_o)\big)$
6, 7	$\phi p_l + (1-\phi)p_h - w$	$\phi p_l + (1-\phi)p_h - w - (1-\phi)\frac{c_u(p_h)}{2}$	$\frac{1}{2}\big(\phi^2 p_l + (1-\phi^2)p_h - w - (1-\phi)c_u(p_h) + \phi c_o\big)$
8	$\phi p_l + (1-\phi)p_h - w$	$\phi p_l + (1-\phi)p_h - w$	$\frac{1}{2}\big(\phi^2 p_l + (1-\phi^2)p_h - w + c_o\big)$

Case k	$\alpha \Upsilon(r)$	$+\frac{\beta_2}{2}\hat{\Lambda}_k(r)$	$-\hat{f}_B\frac{\beta_2}{2}\hat{\Omega}_k(r)$
1, 5, 6	$r - w$	$r - w - c_u(r)$	$r - w - c_u(r)$
2, 3, 4, 7, 8	$r - w$	$r - w$	$r - w + c_o$

In case the competitor's promotion frequency is high, the CHANCE for retailer A to win the smart customer segments β_t is low and consequently, it is optimal for retailer A to order a smaller quantity. This further implies the RISK of high stockout cost for retailer A in case he enters the competition with a high promotion frequency and thereby might succeed in winning the smart customer segments over from his competitor. From his low order quantity, retailer A is not able to serve the smart customer segments in full and he simply cannot afford to promote due to limited stock availability.

Observe that especially in promotions, customers become upset when experiencing the neglected promise of a promotion in stockout situations, and retailers can thus put the loyalty of their customers at stake. Hence, we assume for the stockout cost $g(1 - c(\mathsf{s})) > \mathsf{s} - w$, i.e., the stockout cost g multiplied by the probability of occurrence $1 - c(\mathsf{s})$ at each price level $\mathsf{s} = \{p_l, p_h, r\}$ must be larger than the sales margin $\mathsf{s} - w$.

Consequently, retailer A's best response to a high promotion frequency of retailer B is to stay out of competition and to offer the product at the regular price across all stores, i.e., $\hat{f}_A = 0$. This is the best response for Cases 1 through 7 in order to avoid stockout costs. Hence retailer A can be sure of the profits from the loyal customer segment and speculate on gaining half of the store-switching segment.

Retailer A's best response only changes in Case 8, $\forall \hat{f}_B \in [0, c(p_h)]$, where retailer A provides the full service level to both the loyal and the smart customer segments. In this case, the best response is no longer influenced by the stockout risk, but rather dependent on the size of the stockpiling customer segment. Whenever the size of the stockpiling segment β_{1t} exceeds the critical value τ as characterized in Definition 4.3, the best response of retailer A becomes dependent on the competitor's promotion frequency and retailer A plays strategically. If the stockpiling segment is not large and therefore not profitable enough, i.e., $\beta_{1t} < \tau$, retailer A will again play his pure strategy "regular" price across all stores.

We formulate the best response by employing Definition 4.3, Proposition 4.1 and the procedure described in Sect. 4.2.2.2 in the following proposition.

Proposition 4.4. *$\forall g(1 - c(r)) > r - w$, the best response of retailer A for the scenario of no information sharing $\hat{\delta}_{Ak}(\hat{f}_B, \beta_{1t})$ depends on the size of the stockpiling segment β_{1t} and the competitor's promotion frequency $\hat{f}_B$. The critical size for the stockpiling segment τ is derived as*

$$\tau = \frac{\alpha \Delta \Upsilon - \beta_2 \Delta \hat{\Lambda}_8}{\hat{\Lambda}_8(p)}.$$

We attain two different cases

Case i: *If $\beta_{1t} < \tau$ or $\hat{f}_B \notin [0, c(p_h)]$, then $\hat{\delta}_{Ak}(\hat{f}_B, \beta_{1t}) < 0 \ \forall \hat{f}_B \in [0, 1]$ and retailer A's best response is to play his pure strategy "regular" independent of retailer B's strategy. He has a dominant best response:*

$$\hat{f}_A^* = 0.$$

Case ii: *If $\beta_{1t} < \tau$ and $\hat{f}_B \in [0, c(p_h)]$, retailer A's best response is to play strategically and mix between his pure strategies "regular" and "promotion" with probability*

$$\hat{f}_A^* = \begin{cases} 1 & \text{if } \hat{f}_B \in [0, \hat{f}_B^\dagger), \\ [0,1] & \text{if } \hat{f}_B = \hat{f}_B^\dagger, \\ 0 & \text{if } \hat{f}_B \in (\hat{f}_B^\dagger, 1], \end{cases}$$

with

$$\hat{f}_B^\dagger = \frac{-\Delta\Upsilon + \Delta\hat{\Lambda}_8}{\Delta\hat{\Omega}_8}.$$

Proof. See Appendix A.1 □

The reaction correspondences for retailer A and B for the $k = 8$ cases are plotted in Fig. 4.8.

Observe that the critical size of the stockpiling customer segment τ increases in ϕ:

$$\begin{aligned} \tau &= \frac{\alpha\Delta\Upsilon - \beta_2\Delta\hat{\Lambda}_8}{\hat{\Lambda}_8(p)} \\ &= \frac{\alpha(r - \phi p_l - (1-\phi)p_h) - \beta_2(\phi p_l + (1-\phi)p_h - \frac{r}{2} - \frac{w}{2})}{\phi p_l + (1-\phi)p_h - w}, \\ \frac{\partial\tau}{\partial\phi} &= \frac{\left(\alpha + \frac{\beta_2}{2}\right)(r-w)(p_h - p_l)}{\left(\phi p_l + (1-\phi)p_h - w\right)^2} > 0. \end{aligned} \tag{4.26}$$

As a consequence, the higher the promotion depth of the opponent, the larger the required size of the stockpiling segment β_{1t} to make the retailer respond with a mixed strategy.

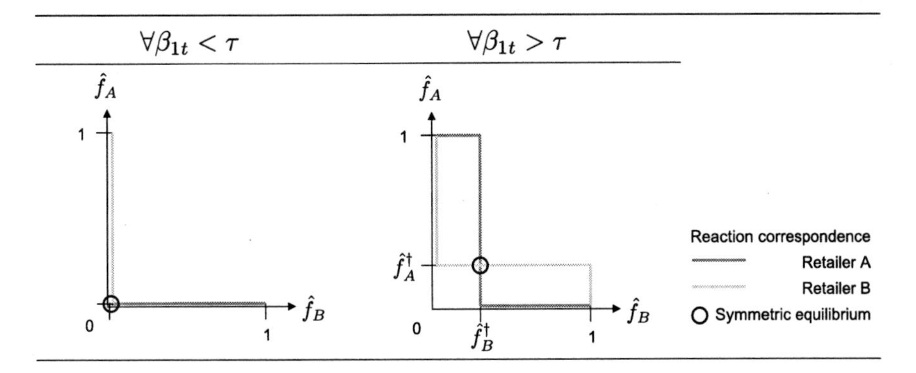

Fig. 4.8 Reaction correspondences for retailer A and B – no information sharing

Symmetric Mixed Strategy Equilibrium

As evident from Proposition 4.3, the Nash equilibria are the points in the intersection of the graphs' of retailer A's and B's best response correspondence as shown in Fig. 4.8. We see that a symmetric Nash equilibrium depends on the size of the switching customer segment and we can observe two different equilibria:

Proposition 4.5. *Under no information sharing, the game has a symmetric Nash equilibrium* $\hat{f}^* = \hat{f}_A^* = \hat{f}_B^*$, *if each retailer plays his best response as described by Proposition 4.1. The equilibrium depends on the size of the stockpiling segment* β_{1t} *and we need to differentiate two cases:*

Case i: *If* $\beta_{1t} < \tau$, *the game has a symmetric pure strategy equilibrium at*

$$\hat{f}^* = 0. \tag{4.27}$$

Consequently both retailers have a dominant strategy "regular".

Case ii: *If* $\beta_{1t} > \tau$, *the game has a symmetric mixed strategy equilibrium at*

$$\hat{f}^* = \frac{-\Delta\Upsilon + \Delta\hat{\Lambda}}{\Delta\hat{\Omega}_8}.$$

Proof. Follows directly from Propositions 4.2 and 4.3. □

An interesting result is that for symmetric retailers, the only equilibrium is found in the full service level case, i.e., Case 8. This result is uncoupled from the size of the stockpiling segment β_{1t}: both the symmetric pure strategy and the symmetric mixed strategy Nash equilibria are within the domain of Case 8. This allows the conclusion that symmetric retailers will always provide the full service level to the customers under no information sharing.

4.3.3 *Information Sharing Scenario*

In this scenario the retailer receives the upstream information from the Competition Index, which provides him either with the information "competitor offers at promotion price" $(\cdot|p)$ or "competitor offers at regular price"$(\cdot|r)$. The retailer receives the information before he places his order with the manufacturer and consequently the information is not yet available at the time the retailer determines his optimal promotion frequency $\check{f}^*$ (compare Sect. 4.1.1).

In the following, we consider the impact of information from the Competition Index on the cumulative distribution function of demand, the optimal order quantity and the retailers' profit. From these, we derive the retailer's best response in order to identify the symmetric mixed strategy equilibrium.

Cumulative Distribution Function of Demand

With information sharing the demand a retailer faces does not change compared to the scenario of no information sharing as shown earlier in Table 4.1. However, the cumulative distribution function $\breve{\Theta}(d(\mathsf{s}_A))$ becomes conditional on the information attained and we need to separate the two cases $\breve{\Theta}(d(\mathsf{s}_A)|\mathsf{p})$ and $\breve{\Theta}(d(\mathsf{s}_A)|r)$ as summarized in Table 4.8 and visualized in Fig. 4.9.

Intuitively, if the retailer attains the information "competitor promotes" the only uncertainty remaining is whether the competitor is promoting at a low p_l or high promotion price p_h, that is there remains uncertainty regarding the promotion depth ϕ. On obtaining the knowledge that the competitor is offering at regular price on the other hand, the uncertainty dissolves and retailer A can be sure to gain the smart customer segments, if he offers at a promotion price, and to serve a share of the store-switching customer segments β_2 if he offers at a regular price.

Table 4.8 Optimal order quantities and cumulative densities for retailer A – information sharing

	Promotion price			Regular price	
	$q_A^*(\mathsf{p}_A)$	$\breve{\Theta}(d(p_l)|\mathsf{p})$	$\breve{\Theta}(d(p_h)|\mathsf{p})$	$q_A^*(r)$	$\breve{\Theta}(d(r)|\mathsf{p})$
Opponent promotion	α		ϕ	α	1
	$\alpha+\frac{\beta_{1t}+\beta_2}{2}$	ϕ	1	$\alpha+\frac{\beta_2}{2}$	1
	$\alpha+\beta_{1t}+\beta_2$	1	1		
	$q_A^*(\mathsf{p}_A)$	$\breve{\Theta}(d(p_l)|r)$	$\breve{\Theta}(d(p_h)|r)$	$q_A^*(r)$	$\breve{\Theta}(d(r)|r)$
Opponent regular	$\alpha+\beta_{1t}+\beta_2$	1	1	$\alpha+\frac{\beta_2}{2}$	1

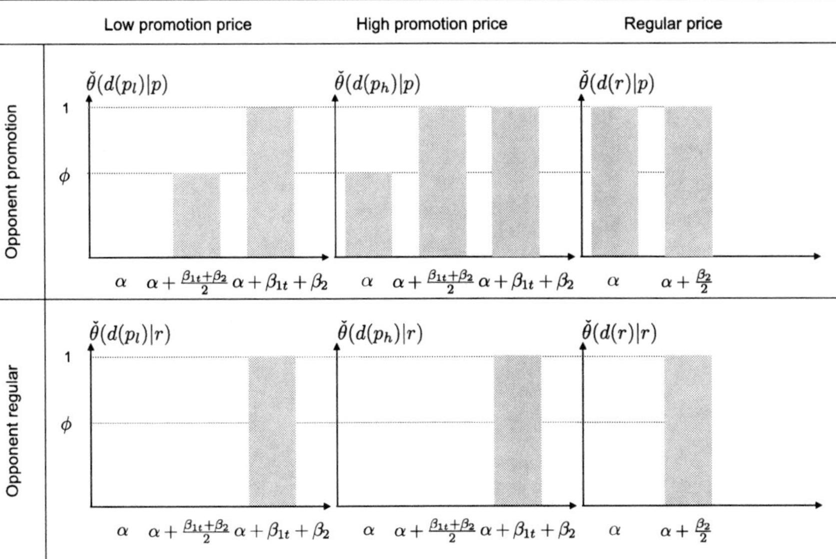

Fig. 4.9 Cumulative distribution function of demand – information sharing

Optimal Order Quantity

Under information sharing, the retailer attains information as to whether his opponent is offering at regular or at promotion price, before he places his order with the manufacturer (compare Sect. 4.1.1). According to the critical fractile (4.3) and the cumulative distribution function of demand $\check{\theta}(d(s_s))$, we attain the following conditional order quantities if retailer A receives the information "opponent promotes":

$$\check{q}_A^*(p_l|p) = \begin{cases} \alpha + \frac{\beta_{1t}+\beta_2}{2} & c(p_l) \in [0,\phi], \\ \alpha + \beta_{1t} + \beta_2 & c(p_l) \in (\phi, 1], \end{cases} \tag{4.28}$$

$$\check{q}_A^*(p_h|p) = \begin{cases} \alpha & c(p_h) \in [0,\phi], \check{f}_B \in [0,1], \\ \alpha + \frac{\beta_{1t}+\beta_2}{2} & c(p_h) \in (\phi, 1], \check{f}_B \in [0,1], \end{cases} \tag{4.29}$$

$$\check{q}_A^*(r|p) = \begin{cases} \alpha \quad c(r) \in [0,1], \check{f}_B \in [0,1]. \end{cases} \tag{4.30}$$

The following conditional order quantities apply, if retailer A receives the information "opponent regular":

$$\check{q}_A^*(p_l|r) = \alpha + \beta_{1t} + \beta_2 \quad c(p_l), \check{f}_B \in [0,1], \tag{4.31}$$

$$\check{q}_A^*(p_h|r) = \alpha + \beta_{1t} + \beta_2 \quad c(p_h), \check{f}_B \in [0,1], \tag{4.32}$$

$$\check{q}_A^*(r|r) = \alpha + \frac{\beta_2}{2} \quad c(r), \check{f}_B \in [0,1]. \tag{4.33}$$

If we combine the conditional order quantities at the respective price levels, we attain, according to Definition 4.1, $k = 3$ different cases for the information sharing scenario. The combination of conditional optimal order quantities at the different price levels is summarized in Table 4.9. In Fig. 4.10, we plot the promotion depth ϕ along the x-axis with its bounds $0, c(p_l), c(p_h)$ and 1. The y-axis is the competitor's promotion frequency $\check{f}_B$ in its domain $\check{f}_B \in [0,1]$. The resulting areas for the $k = 3$ cases are also indicated.

The optimal order quantity decreases in the competitor's promotion depth ϕ: the higher the promotion depth of the opponent, the lower the probability of attaining the smart customer segments and consequently the lower the service level, i.e., the

Table 4.9 Domains for the promotion frequencies $\check{f}_B \in [0,1]$ and ϕ and the respective optimal order quantities

Case	$\phi \in$	$\check{q}^*(p_l\|p)$	$\check{q}^*(p_h\|p)$	$\check{q}^*(r\|p)$	$\check{q}^*(p_l\|r)$	$\check{q}^*(p_h\|r)$	$\check{q}^*(r\|r)$
1	$[c(p_h), 1]$	$\alpha + \frac{\beta_{1t}+\beta_2}{2}$	α	α	$\alpha + \beta_{1t} + \beta_2$	$\alpha + \beta_{1t} + \beta_2$	$\alpha + \frac{\beta_2}{2}$
2	$[c(p_l), c(p_h)]$	$\alpha + \frac{\beta_{1t}+\beta_2}{2}$	$\alpha + \frac{\beta_{1t}+\beta_2}{2}$	α	$\alpha + \beta_{1t} + \beta_2$	$\alpha + \beta_{1t} + \beta_2$	$\alpha + \frac{\beta_2}{2}$
3	$[0, c(p_l)]$	$\alpha + \beta_{1t} + \beta_2$	$\alpha + \frac{\beta_{1t}+\beta_2}{2}$	α	$\alpha + \beta_{1t} + \beta_2$	$\alpha + \beta_{1t} + \beta_2$	$\alpha + \frac{\beta_2}{2}$

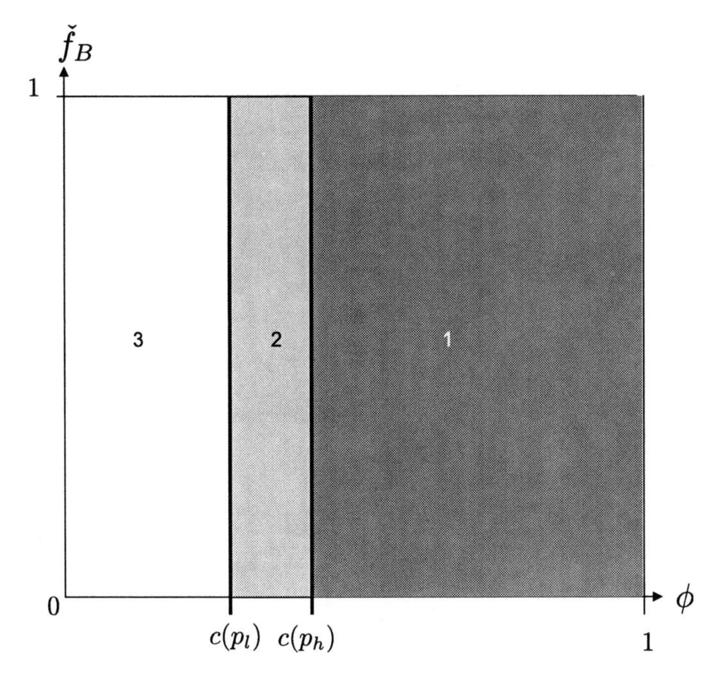

Fig. 4.10 Domain of cases for the optimal order quantity – information sharing

lower the optimal order quantity. In Case 3, the optimal order quantity of retailer A is highest, providing full service level to all three customer segments α, β_{1t} and β_2 across all three price levels p_l, p_h and r. It is lowest in Case 2, where the full service level is only provided in case the retailer attains the information that the competitor is offering at regular price.

Expected Profit Under Mixed Strategies

With information sharing, the optimal order quantity defines the domain of the mixing probability ϕ, which are characterized in Table 4.9. Within these domains, we determine the expected profit under mixed strategies from (4.4), (4.6) and (4.10) for each of the $k = 3$ cases as

$$\check{\pi}_{Ak}(\check{f}_A; \check{f}_B, \phi) = \check{f}_A \check{\pi}_{Ak}(\mathsf{p}; \check{f}_B, \phi) + (1 - \check{f}_A)\check{\pi}_{Ak}(r; \check{f}_B, \phi) \tag{4.34}$$

with

$$\check{\pi}_{Ak}(\mathsf{p}; \check{f}_B, \phi) = \Upsilon(\mathsf{p}) + \check{\Lambda}_k(\mathsf{p}) - \check{f}_B \check{\Omega}_k(\mathsf{p}) \tag{4.35}$$

and

$$\pi_{Ak}(r; \check{f}_B) = \Upsilon(r) + \check{\Lambda}_k(r) - \check{f}_B \check{\Omega}_k(r), \tag{4.36}$$

where

$$\begin{aligned}
\check{\Upsilon}(\mathsf{p}) &= \sigma\alpha(\mathsf{p} - w),\\
\check{\Upsilon}(r) &= \alpha(r - w),\\
\check{\Lambda}(\mathsf{p}) &= \sigma(\beta_{1t} + \beta_2)(\mathsf{p} - w),\\
\check{\Lambda}(r) &= \frac{\beta_2}{2}(r - w),\\
\check{\Omega}_k(\mathsf{p}) &= \sigma\chi\frac{\beta_{1t} + \beta_2}{2}(\mathsf{p} - w) + \sigma\check{\Gamma}_k(\mathsf{p}),\\
\check{\Omega}_k(r) &= \frac{\beta_2}{2}(r - w).
\end{aligned}$$

Finally,

$$\begin{aligned}
\check{\Gamma}_k(\mathsf{p}) = c_u(\mathsf{p})\big((1-\phi)\max[d(\mathsf{p}, p_h) - q_k^*(\mathsf{p}|p), 0]\big)\\
+ c_o\big(\phi\max[q_k^*(\mathsf{p}|p) - d(\mathsf{p}, p_l), 0]\big).
\end{aligned}$$

Inventory cost can only occur in case retailer A receives the information "opponent promotes" due to the uncertainty about the competitor's promotion depth ϕ, that is the uncertainty as to whether the competitor is promoting at the high promotion price p_h or at the low promotion price p_l. On the contrary, uncertainty resolves if retailer A is aware that the opponent is offering at regular price r and intuitively inventory costs diminish.

From $\check{\Lambda}(\mathsf{p})$ and $\check{\Lambda}(r)$ it becomes evident that with information sharing both the promotion CHANCE and the regular CHANCE are positive, independent of the order quantity. Whenever the competitor offers at regular price, it is known to retailer A. Hence, he does not incur inventory costs as in the scenario of no information sharing. Both the promotion CHANCE and the regular CHANCE are the same for all three scenarios. Under these premises, the promotion CHANCE $\check{\Lambda}(p)$ and the regular CHANCE $\check{\Lambda}(r)$ do not change for the $k = 3$ cases, allowing us to suppress the index k.

The profits for the three cases are summarized in Table 4.10.

Best Response

From (4.15), we can derive the best response on information sharing as

$$\check{\delta}_{Ak}(\check{f}_B, \beta_{1t}) = -\Delta\Upsilon + \Delta\check{\Lambda} - \check{f}_B\Delta\check{\Omega}_k. \tag{4.37}$$

Under information sharing, the BASE and CHANCE elements of the best response are positive, if the size of the stockpiling segment β_{1t} is large enough, that is

$$\begin{aligned}
\check{\delta}_{Ak}(\check{f}_B = 0) = -\Delta\Upsilon + \Delta\Lambda > 0,\\
\beta_{1t} > \frac{\alpha\Delta\Upsilon - \beta_2\Delta\Lambda}{\Lambda(p)},
\end{aligned}$$

Table 4.10 Overview of profit components – information sharing

Case k	BASE	CHANCE	RISK
	$\alpha\Upsilon(p)$	$+(\beta_{1t}+\beta_2)\check{\Lambda}_k(p)$	$-\check{f}_B(\beta_{1t}+\beta_2)\check{\Omega}_k(p)$
1	$\phi p_l+(1-\phi)p_h-w$	$\phi p_l+(1-\phi)p_h-w$	$\frac{1}{2}\left(\phi^2 p_l+(1-\phi^2)p_h-w+\phi(1-\phi)c_u(p_l)+(1-\phi)^2 c_u(p_h)\right)$
2	$\phi p_l+(1-\phi)p_h-w$	$\phi p_l+(1-\phi)p_h-w$	$\frac{1}{2}\left(\phi^2 p_l+(1-\phi^2)p_h-w+\phi(1-\phi)(c_u(p_l)+c_o)\right)$
3	$\phi p_l+(1-\phi)p_h-w$	$\phi p_l+(1-\phi)p_h-w$	$\frac{1}{2}\left(\phi^2 p_l+(1-\phi^2)p_h-w+\phi c_o\right)$

Case k	$\alpha\Upsilon(r)$	$+\frac{\beta_2}{2}\check{\Lambda}_k(r)$	$-\check{f}_B\frac{\beta_2}{2}\check{\Omega}_k(r)$
1, 2, 3	$r-w$	$r-w$	$r-w$

which is described by the critical size τ of the stockpiling customer segment in Definition 4.3. Hence, whenever the stockpiling segment exceeds its critical size, it is optimal for the retailer to play a mixed strategy. In contrast, if the segment is too small and therefore unprofitable, the retailer's best response is to play his regular price and base his profits on the loyal customer segment. We summarize the results for the best response in the following proposition.

Proposition 4.6. *With information sharing, the best response $\check{\delta}_{Ak}(\check{f}_B, \beta_{1t})$ of retailer A depends on the size of the stockpiling segment β_{1t} and the competitor's promotion frequency $\check{f}_B$. The critical size for the stockpiling segment τ is derived as*

$$\begin{aligned}\tau &= \frac{\alpha \Delta \Upsilon - \beta_2 \Delta \Lambda}{\Lambda(p)} \\ &= \frac{\alpha(r - \phi p_l - (1-\phi) p_h) - \beta_2(\phi p_l + (1-\phi) p_h - \frac{r}{2} - \frac{w}{2})}{\phi p_l + (1-\phi) p_h - w}\end{aligned}$$

and we can describe the following two cases:

Case i: *If $\beta_{1t} < \tau$, then $\check{\delta}_{Ak}(\check{f}_B, \beta_{1t}) < 0 \; \forall \check{f}_B \in [0,1]$ and retailer A's best response is to play his pure strategy "regular" no matter how retailer B plays: He has a dominant best response.*

$$\check{f}_A^* = 0.$$

Case ii: *If $\beta_{1t} > \tau$, retailer A's best response is to play strategically and mix between his pure strategies "regular" and "promotion" with probability*

$$\check{f}_A^* = \begin{cases} 1 & \text{if } \check{f}_B \in [0, \check{f}_B^\dagger), \\ [0,1] & \text{if } \check{f}_B = \check{f}_B^\dagger, \\ 0 & \text{if } \check{f}_B \in (\check{f}_B^\dagger, 1], \end{cases}$$

with

$$\check{f}_B^\dagger = \frac{-\Delta \Upsilon + \Delta \check{\Lambda}}{\Delta \check{\Omega}_k}.$$

Proof. See Appendix A.2. □

In contrast to the scenario of no information sharing, it can be optimal for the retailer to order below the full service level in the information sharing scenario (Case 1 and Case 2).

The reaction correspondences for retailer A and retailer B for the $k = 3$ cases of the information sharing scenario are visualized in Fig. 4.11.

Observe that the critical size τ of the stockpiling segment is the same as in the no information sharing scenario. Consequently, the results attained in (4.26) also apply to this scenario: τ increases along with the promotion depth ϕ, which implies that

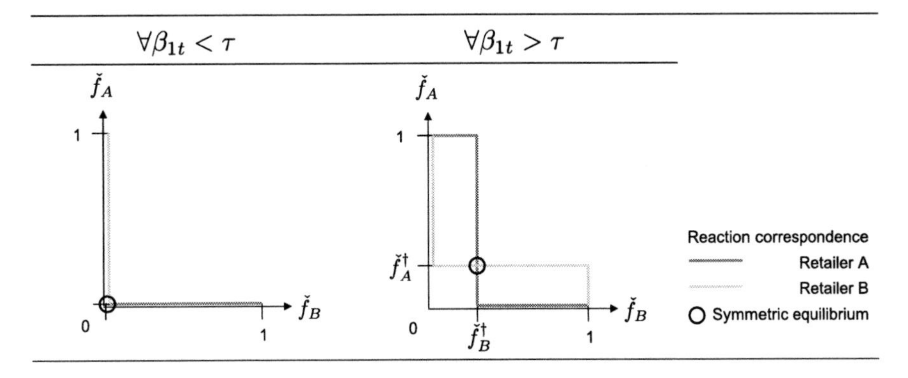

Fig. 4.11 Reaction correspondences for retailer A and B – information sharing

the more frequently the retailer offers the product at the low promotion price p_l, the larger the required size of the stockpiling segment β_{1t} in order to make the retailer respond with a mixed strategy.

Symmetric Mixed Strategy Equilibrium

With variations of the size of the stockpiling customer segment, we show that the game is not only solved in mixed strategy but also in pure strategy Nash equilibria. Under a Nash equilibrium, each retailer plays his best response. Hence, the Nash equilibrium is completely characterized by the best response as defined in Proposition 4.2 and found at the intersection of the best responses of retailer A and B in Fig. 4.11, and we can summarize the symmetric mixed strategy Nash equilibrium in the following proposition:

Proposition 4.7. *With information sharing, the game has a symmetric Nash equilibrium $\check{f}^* = \check{f}_A^* = \check{f}_B^*$, if each retailer plays his best response as described by Proposition 4.6. The equilibrium depends on the stockpiling segment β_{1t} and two different symmetric Nash equilibria can emerge:*

Case i: *$\forall \beta_{1t} < \tau$ a symmetric pure strategy equilibrium exists with*

$$\check{f} = 0. \tag{4.38}$$

That is, both retailers have the dominant strategy "regular".

Case ii: *$\forall \beta_{1t} > \tau$ a symmetric mixed strategy equilibrium exists at*

$$\check{f}^* = \frac{-\Delta \Upsilon + \Delta \check{\Lambda}}{\Delta \check{\Omega}_k}.$$

Proof. Follows directly from Propositions 4.2 and 4.3. □

Observe that with information sharing, the order quantity provided by the symmetric retailers depends purely on the exogenously set promotion depth ϕ. This result is in contrast to the result of the no information sharing scenario where the order quantity is subject to the opponents promotion frequency as well. Whenever ϕ is high, that is there is a high probability of the low promotion price p_l, the retailers order a smaller quantity from the manufacturer, than for larger ϕ (compare Table 4.9).

4.4 The Value of Upstream Information Sharing

Receiving the upstream information from the Competition Index allows the retailer to conclude on whether his competitor is offering the product at the promotion or at the regular price. Inherently, by reducing the pricing uncertainty, the demand uncertainty of retailers is reduced. This in turn allows the retailer to improve his order decisions resulting in lower inventory cost. But to what extent do retailers and customers benefit from the information?

The impact of information sharing for the retailer can be broken down into three different effects: inventory effect, frequency effect and profit effect. The inventory effect describes the impact of information sharing on the inventory cost of a promotion. If the cost of a promotion decreases, the frequency with which a promotion is conducted increases, resulting in the frequency effect. Further, if both the inventory and the frequency effect are positive, the profit should be higher with information sharing than with no information sharing for the retailer. We shall provide evidence that the retailer can increase his profits with information sharing.

Further, we consider the impact of upstream information sharing on customer welfare, individually for loyal customers, stockpiling customers and store-switching customers. We will find a positive welfare effect for all three customer segments. Consequently, the retailer benefits from information sharing are not borne by the customer, but the retailer benefit from the result of decreased inventory cost and hence an increased supply chain efficiency.

The magnitude of effects, i.e., the profit increase for the retailer and the welfare increase for the customers, depends on the size of the three customer segments. As evident from the results of the previous chapter, the more stockpilers are in the market, the higher the benefits from information sharing for the retailer and the higher the welfare effect for the customer.

To analyze the value of information, we compare all cases of the two scenarios, information sharing and no information sharing, where an equilibrium can occur. According to Proposition 4.5 for the no information sharing scenario and Proposition 4.7 for the information sharing scenario, we hence need to compare Case 8 for the no information sharing scenario with the three different cases of information sharing. An overview on the domain of ϕ for the comparison is provided in Table 4.11.

Table 4.11 Relevant cases for comparison

No information sharing	Information sharing		
Case 8	Case 1	Case 2	Case 3
$\phi \in [0, 1]$	$\phi \in [c(p_h), 1]$	$\phi \in [c(p_l), c(p_h)]$	$\phi \in [0, c(p_l)]$

4.4.1 Inventory Effect

With information sharing, the retailer attains information about the competitor's pricing strategy which in turn reduces the demand uncertainty. The question is, how does the diminishing demand uncertainty impact the inventory cost?

In order to derive an answer to this question, we assume constant promotion frequencies $\check{f} = \hat{f}$ and extract the difference in inventory cost between the two scenarios. The comparison of profits of the two scenarios, information sharing (4.34) and no information sharing (4.22) reveals

$$\begin{aligned}
\Delta\pi_k &= \check{\pi}_k(f) - \hat{\pi}_8(f) \\
&= \beta_{1t} f_A f_B \Big(\hat{\Omega}_8(p) - \check{\Omega}_k(p) \Big) \\
&+ \beta_2 f_B \bigg(\frac{1}{2} (\hat{\Omega}_8(r) - \check{\Omega}_k(r)) \\
&+ f_A \left(\hat{\Omega}_8(p) - \check{\Omega}_k(p) - \frac{1}{2} (\hat{\Omega}_8(r) + \check{\Omega}_k(r)) \right) \bigg) \\
&= \beta_{1t} f_A f_B \sigma (\check{\Gamma}_k(p) - \hat{\Gamma}_8(p)) \\
&+ \beta_2 f_B \sigma \left(-\frac{1}{2} \hat{\Gamma}_8(r) + f_A \left(\check{\Gamma}_k(p) - \hat{\Gamma}_8(p) + \frac{1}{2} \hat{\Gamma}_8(r) \right) \right).
\end{aligned} \tag{4.39}$$

The difference in profits simplifies to a pure inventory problem. We denote this effect as inventory effect $\Delta\pi_k = \check{\pi}_k(f) - \hat{\pi}_8(f)$, $k = 1, \ldots, 3$. The inventory effect describes the impact of information sharing on the inventory cost for constant frequencies $\check{f} = \hat{f}$. The effect is formalized in the following proposition.

Proposition 4.8. *The inventory cost associated with promotions is reduced with information sharing as compared to no information sharing, if the promotion frequency f is kept constant across scenarios. Formally* $\Delta\pi_k = \check{\pi}_k(f) - \hat{\pi}_8(f) \geq 0$, $\forall \phi \in [0, 1]$ *and* $f \in (0, 1]$.

Proof. We analyze each domain of ϕ in turn:

1. $\phi \in [c(p_h), 1]$ Case 1 Information sharing vs. Case 8 No information sharing:

$$\Delta\pi_1 = \frac{\beta_{1t}}{2} f^2\big(c_o - (1-\phi)(\phi c_u(p_l) + (1-\phi)c_u(p_h))\big) + \frac{\beta_2}{2} f\big(c_o - f(1-\phi)(\phi c_u(p_l) + (1-\phi)c_u(p_h))\big).$$

We proceed by showing that the inventory cost associated with β_{1t} and β_2 are positive terms. If the term associated with β_{1t}, $c_o - (1-\phi)(\phi c_u(p_l) + (1-\phi)c_u(p_h)) > 0$, so is the β_2-term, because $0 < f \leq 1$ and we can focus on

$$c_o - (1-\phi)(\phi c_u(p_l) + (1-\phi)c_u(p_h)) > 0.$$

We approximate $\phi c_u(p_l) + (1-\phi)c_u(p_h) \approx c_u(p_h)$ without loss of generality, because $c_u(p_h)$ is the larger and consequently more critical value of the two. Then,

$$c_o - (1-\phi)c_u(p_h) > 0,$$
$$\phi > 1 - \frac{c_o}{c_u(p_h)}.$$

Given the domain of $\phi \in [c(p_h), 1]$, we need to show that

$$\begin{aligned} \phi = c(p_h), \quad & c(p_h) > 1 - \tfrac{c_o}{c_u(p_h)}, \\ & \tfrac{c_u(p_h)}{c_u(p_h)+c_o} > 1 - \tfrac{c_o}{c_u(p_h)}, \\ & 0 > -c_o^2, \\ \phi = 1, \quad & 1 > 1 - \tfrac{c_o}{c_u(p_h)}, \end{aligned}$$

which is always fulfilled. Thus, we attain a positive inventory effect $\forall \phi \in [c(p_h), 1]$.

2. $\phi \in [c(p_l), c(p_h)]$ Case 2 Information sharing vs. Case 8 No information sharing:

$$\Delta\pi_2 = \frac{\beta_{1t}}{2} f^2\big(c_o - \phi(1-\phi)(c_u(p_l) + c_o)\big) + \frac{\beta_2}{2} f\big(c_o - f\phi(1-\phi)(c_u(p_l) + c_o)\big).$$

In order to proof that the inventory effect is positive, we need to show that the terms associated with β_{1t} and β_2 are positive. Again, we can restrict our attention to the term associated with β_{1t}, $c_o - \phi(1-\phi)(c_u(p_l) + c_o) > 0$, because it is the smaller one due to $0 < f \leq 1$.

The term $c_o - \phi(1-\phi)(c_u(p_l)+c_o)$ has its minimum at $\phi = \frac{1}{2}$ and we can show that even at the minimum, the term is still positive:

$$\phi = \frac{1}{2} \Rightarrow c_o - \frac{1}{4}(c_u(p_l) + c_o) > 0,$$
$$\frac{c_o}{c_u(p_l) + c_o} > \frac{1}{4},$$
$$1 - c(p_l) > \frac{1}{4},$$
$$c(p_l) < \frac{3}{4}.$$

Hence, we attain a positive profit effect if the underage cost $c(p_l) < \frac{3}{4}$. In case the underage cost exceeds this value, we require the size of the stockpiling segment to be

$$\beta_{1t} < \beta_2 \frac{c_o - f\phi(1-\phi)(c_u(p_l) + c_o)}{c_o - \phi(1-\phi)(c_u(p_l) + c_o)},$$

which is always fulfilled, and we thus attain a positive inventory effect $\forall \phi \in [c(p_l), c(p_h)]$.

3. $\phi \in [0, c(p_l)]$ Case 3 Information sharing vs. Case 8 No information sharing:

$$\Delta\pi_3 = \left(\frac{\beta_{1t}}{2} f^2(1-\phi) + \frac{\beta_2}{2} f(1 - f\phi)\right) \frac{c_o}{2} > 0,$$

which is always fulfilled. Thus, we attain a positive inventory effect $\forall \phi \in [0, c(p_l)]$. □

Intuitively, the retailer has lower inventory costs with information sharing: if the retailer is aware of the opponents pricing strategy, the demand uncertainty is reduced, resulting in a better match of orders to actual demand which in turn reduces inventory cost. Further, the inventory effect is largest for $\phi \in [c(p_h), 1]$, i.e., Case 8 no information sharing and Case 1 information sharing. Within this domain, the optimal order quantity of the retailer is the smallest among the three cases with the largest difference to the high order quantity in information sharing.

When looking at the inventory effect by customer segments, it turns out that only the loyal customer segment α has no impact. The loyal customer segment returns to the store every week, creating a constant and certain demand. Consequently, no inventory cost occurs for the loyal customer segment in either scenario.

Further, the inventory effect is larger for the store-switching segment β_2 than for the stockpiling segment β_{1t}. This becomes evident when comparing the terms associated with β_{1t} and β_2 from (4.39) as

$$f_A f_B \sigma(\check{\Gamma}_k(p) - \hat{\Gamma}_8(p)) < f_B \sigma \left(-\frac{1}{2}\hat{\Gamma}_8(r) + f_A \left(\check{\Gamma}_k(p) - \hat{\Gamma}_8(p) + \frac{1}{2}\hat{\Gamma}_8(r)\right)\right),$$

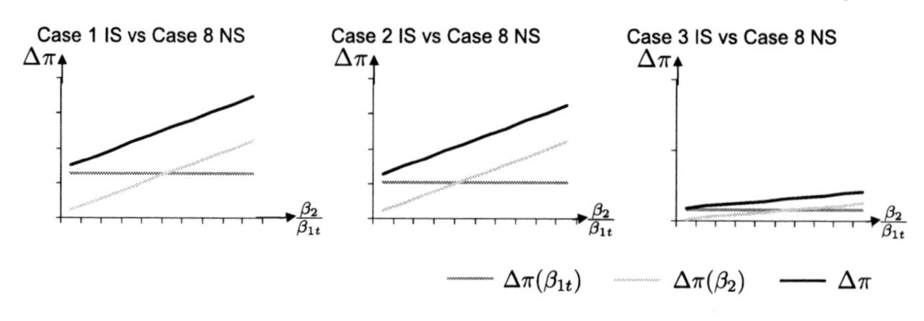

Fig. 4.12 Inventory effect of smart customer segments

which is fulfilled, given that Γ_i is defined as negative as given in (4.5). Hence, the larger the size of the store-switching segment β_2 relative to the stockpiling segment β_{1t}, the larger the inventory effect. The difference for the three cases is summarized in Fig. 4.12 where we show the total inventory effect $\Delta\pi_k$, the inventory effect from the stockpiling segment $\Delta\pi_k(\beta_{1t})$ and from the store-switching segment $\Delta\pi_k(\beta_2)$ for an increasing ratio of $\frac{\beta_2}{\beta_{1t}}$, while keeping β_{1t} constant.

4.4.2 Frequency Effect

In the previous section, we have shown that the inventory costs associated with promotions are reduced with information sharing. If promotions are less expensive, we expect the retailer to promote more frequently in equilibrium. In order to find evidence for an increasing frequency with information sharing – the frequency effect Δf^* – we compare the equilibrium promotion frequencies f^* of the two scenarios as given by Proposition 4.5, for the scenario of no information sharing and Proposition 4.7 for the scenario of information sharing.

The equilibrium promotion frequency f^* in either scenario depends on the size of the stockpiling segment β_{1t}: if the segment is smaller than its critical size τ, the retailer's equilibrium promotion frequency is $f^* = 0$ in both scenarios. Consequently, the frequency remains constant at zero across scenarios and information sharing has no impact on the promotion frequency. However, for any size of the stockpiling segment β_{1t} exceeding τ, the retailer promotes. He promotes at $\hat{f}_8^*$ in the scenario of no information sharing and at $\check{f}_k^*$ in the scenario of no information sharing:

$$\Delta f^* = \check{f}_k^* - \hat{f}_8^* \geq 0,$$

$$\forall \beta_{1t} < \tau,\ \Delta f^* = 0,$$

$$\forall \beta_{1t} > \tau,\ \frac{\Delta\Upsilon + \Delta\Lambda}{\Delta\check{\Omega}_k} - \frac{\Delta\Upsilon\Delta\Lambda}{\Delta\hat{\Omega}_8} > 0.$$

If $\beta_{1t} > \tau$, the denominators of the promotion frequencies on information sharing and no information sharing are equal. With this established, neither in the case of full service level with no information sharing nor in any case of information sharing, will underage cost occur if the competitor offers at the regular price. Therefore, the comparison of the cases reduces to a comparison of RISKS on information sharing versus no information sharing.

$$\Delta\hat{\Omega}_8 > \Delta\check{\Omega}_k,$$

which can be rearranged to

$$\beta_{1t}\sigma\big(\check{\Gamma}_k(p) - \hat{\Gamma}_8(p)\big) + \beta_2\sigma\left(-\frac{1}{2}\hat{\Gamma}_8(r) + \left(\check{\Gamma}_k(p) - \hat{\Gamma}_8(p) + \frac{1}{2}\hat{\Gamma}_8(r)\right)\right) > 0$$

implying that the RISK of serving the β_{1t} and the β_2 customer segment must be reduced in information sharing in order to attain a positive frequency effect. We use the following proposition to characterize the frequency effect.

Proposition 4.9. *If $\beta_{1t} > \tau$ the equilibrium promotion frequency increases in information sharing as compared to in no information sharing, i.e., $\Delta f^* = \check{f}^* - \hat{f}^* > 0$. If $\beta_{1t} < \tau$ retailers have the same equilibrium promotion frequency with and without information sharing, i.e., $\Delta f^* = 0$. $\forall\phi \in [0,1]$.*

Proof. If $\beta_{1t} < \tau$, then according to Propositions 4.5 and 4.7, the equilibrium promotion frequency in both scenarios is $f^* = 0$ and consequently, the frequency effect $\Delta f^* = 0$.

If $\beta_{1t} > \tau$, we need to compare the three cases from the information sharing scenario to Case 8 of the no information sharing scenario, as summarized in Table 4.11, individually.

However, we will first make an important note, which holds true for all three cases: If $\beta_{1t} > \tau$, β_{1t} is significantly larger than β_2 given the assumption $\alpha > \beta_2$:

$$\begin{aligned}\beta_{1t} > \tau &= \frac{\alpha\Delta\Upsilon - \beta_2\Delta\Lambda}{\Lambda(p)}\\ &= \frac{\alpha(r - \phi p_l - (1-\phi)p_h) + \beta_2(\frac{r}{2} + \frac{w}{2} - \phi p_l - (1-\phi)p_h)}{\phi p_l + (1-\phi)p_h - w}.\end{aligned}$$

In order to show that $\Delta f^* > 0$, $\forall\phi \in [0,1]$, $\beta_{1t} > \tau$, we now consider the domains individually:

1. $\phi \in [c(p_h), 1]$ Case 1 Information sharing vs. Case 8 No information sharing:

$$\begin{aligned}&\frac{\beta_{1t}}{2}\big(c_o - (1-\phi)\big(\phi c_u(p_l) + (1-\phi)c_u(p_h)\big)\big)\\ &\qquad > \frac{\beta_2}{2}(1-\phi)\big(\phi c_u(p_l) + (1-\phi)c_u(p_h)\big),\\ &\frac{c_o}{(1-\phi)\big(\phi c_u(p_l) + (1-\phi)c_u(p_h)\big)} - 1 > \frac{\beta_2}{\beta_{1t}}.\end{aligned}$$

We approximate $\phi c_u(p_l) + (1-\phi)c_u(p_h) \approx c_u(p_h)$ without loss of generality, given that $c_u(p_h)$ is the larger and consequently the more critical term

$$\frac{c_o}{(1-\phi)c_u(p_h)} - 1 > \frac{\beta_{1t}}{\beta_2},$$
$$\frac{c_o}{c_u(p_h)}\frac{(c_u(p_h)+c_o)}{(c_u(p_h)+c_o)} > (1-\phi)\left(\frac{1+\beta_{1t}}{\beta_2}\right),$$
$$\frac{1-c(p_h)}{c(p_h)} > (1-\phi)\left(1+\frac{\beta_{1t}}{\beta_2}\right),$$
$$\phi > 1 - \frac{1-c(p_h)}{c(p_h)\left(1+\frac{\beta_{1t}}{\beta_2}\right)}.$$

We confirm that the condition holds true at either bound of the domain of $\phi \in [c(p_h), 1]$, and therefore the frequency effect will also exist for all interior values

$$\phi = c(p_h),\ c(p_h) > 1 - \frac{1-c(p_h)}{c(p_h)\left(1+\frac{\beta_{1t}}{\beta_2}\right)},$$
$$\frac{\beta_{1t}}{\beta_2} < \frac{1-c(p_h)}{c(p_h)},$$
$$\phi = 1,\ 1 > 1 - \frac{1-c(p_h)}{c(p_h)\left(1+\frac{\beta_{1t}}{\beta_2}\right)},$$
$$0 > -\frac{1-c(p_h)}{c(p_h)\left(1+\frac{\beta_{1t}}{\beta_2}\right)}.$$

For $\phi = 1$ the frequency effect is positive. At $\phi = c(p_h)$, an additional condition for the ratio of the stockpiling segment and the store-switching segment is required

$$\frac{\beta_{1t}}{\beta_2} < \frac{1-c(p_h)}{c(p_h)},$$

which holds true, given that $\beta_{1t} > \tau$ and hence

$$\beta_{1t} > \beta_2 \frac{c(p_h)}{1-c(p_h)}.$$

Consequently, we attain a positive frequency effect $\forall \phi \in [c(p_h), 1]$.

2. $\phi \in [c(p_l), c(p_h)]$ Case 2 Information sharing vs. Case 8 No information sharing:

$$\frac{\beta_{1t}}{2}\big(c_o - \phi(1-\phi)\big(c_u(p_l)+c_o\big)\big) > \frac{\beta_2}{2}\phi(1-\phi)\big(c_u(p_l)+c_o\big),$$
$$\frac{c_o}{\phi(1-\phi)\big(c_u(p_l)+c_o\big)} - 1 > \frac{\beta_2}{\beta_{1t}},$$
$$1 - c(p_l) > \phi(1-\phi)\big(1+\frac{\beta_2}{\beta_{1t}}\big).$$

If the equation holds at either bound of the domain of $\phi \in [c(p_l), c(p_h)]$, the frequency effect is also positive for all interior values

$$\phi = c(p_l),\ (1-c(p_l)) > c(p_l)(1-c(p_l))\big(1+\frac{\beta_2}{\beta_{1t}}\big),$$
$$\frac{\beta_2}{\beta_{1t}} < \frac{1-c(p_l)}{c(p_l)},$$
$$\phi = c(p_h),\ (1-c(p_l)) > c(p_h)(1-c(p_h))\big(1+\frac{\beta_2}{\beta_{1t}}\big),$$
$$\approx c(p_h)(1-c(p_l))\big(1+\frac{\beta_2}{\beta_{1t}}\big),$$
$$\frac{\beta_2}{\beta_{1t}} < \frac{1-c(p_h)}{c(p_h)}.$$

Again, we attain a requirement for the ratio of stockpilers vs. store switchers as $\frac{\beta_2}{\beta_{1t}} < \frac{1-c(p_h)}{c(p_h)}$ and $\frac{\beta_2}{\beta_{1t}} < \frac{1-c(p_l)}{c(p_l)}$ at either bound, where we can focus on the first requirement, because

$$\frac{1-c(p_h)}{c(p_h)} < \frac{1-c(p_l)}{c(p_l)},$$
$$c(p_l) < c(p_h).$$

And by the same argument as before, we attain a positive frequency effect $\forall\phi \in [c(p_l), c(p_h)]$.

3. $\phi \in [0, c(p_l)]$ Case 3 Information sharing vs. Case 8 No information sharing:

$$\frac{\beta_{1t}}{2}(1-\phi)c_o > \frac{\beta_2}{2}\phi c_o,$$
$$\frac{1-\phi}{\phi} > \frac{\beta_2}{\beta_{1t}}.$$

Again, we confirm that the frequency effect is positive at the bounds $\phi \in [0, c(p_l)]$

$$\phi = 0, \qquad 1 > 0,$$
$$\phi = c(p_l),\ \frac{\beta_2}{\beta_{1t}} < \frac{1-c(p_l)}{c(p_l)},$$

and find the familiar requirement for the ratio of stockpilers versus store switchers, which allows us to conclude that the frequency effect is also positive $\forall \phi \in [0, c(p_l)]$. □

When analyzing the impact of the customer segments on the frequency effect, we do not find any influence from the loyal customer segment α. The smart customer segments β_t can be attracted by promotions at lower inventory cost (positive inventory effect), making it profitable for the retailer to employ promotions more frequently. Among the smart customer segments, the stockpiling segment β_{1t} has a larger impact than the store-switching segment β_2. Competing for the stockpiling segment incentivises the retailer to promote. Naturally, the more stockpilers are in the market, the larger the retailer's equilibrium promotion frequency. With the positive inventory effect in information sharing, the inventory cost of a promotion are reduced. Consequently, the more stockpilers are in the market, the larger the frequency effect.

4.4.3 *Profit Effect*

Given the results from the inventory effect and the frequency effect, we now focus on the profit effect in order to determine the impact of information sharing on the retailer's profit under equilibrium strategies.

The profit under equilibrium promotion frequency for symmetric retailers is defined in Sect. 4.2.2.3 as

$$\beta_{1t} < \tau, \quad \pi_{Ak}(0, 0, \phi) = \left(\alpha + \frac{\beta_2}{2}\right)(r - w), \tag{4.40}$$

$$\begin{aligned} \beta_{1t} > \tau, \quad \pi_{Ak}(f_A^*, f_B^*, \phi) &= f_A^* \pi_{Ak}(p) + (1 - f_A^*)\pi_{Ak}(r) \\ &= \Upsilon(r) + \Lambda_k(r) - f_B^* \Omega_k(r). \end{aligned} \tag{4.41}$$

Then the difference of the profits in equilibrium for the no information sharing and information sharing scenarios reduces for symmetric retailers $f^* = f_A^* = f_B^*$ to

$$\begin{aligned} \Delta\pi_k^* &= \check{\pi}_{Ak}(\check{f}_A^*, \check{f}_B^*, \phi) - \hat{\pi}_{Ak}(\hat{f}_A^*, \hat{f}_B^*, \phi) \\ &= \frac{\beta_2}{2}\left(\hat{f}^* \hat{\Omega}_8(r) - \check{f}^* \check{\Omega}_k(r)\right). \end{aligned}$$

We denote the profit effect as $\Delta\pi_k^*$ and formulate the following proposition to characterize the profit effect.

Proposition 4.10. *If $\beta_{1t} > \tau$, symmetric retailers make a higher profit with information sharing than with no information sharing in equilibrium, i.e., $\Delta\pi_k^* > 0$. If $\beta_{1t} < \tau$, retailers make the same equilibrium profit with and without information sharing, i.e., $\Delta\pi_k^* = 0$. The result $\forall \phi \in [0, 1]$ holds true.*

Proof. If $\beta_{1t} < \tau$, then the retailer's equilibrium promotion frequency $f^* = 0$ for the information sharing scenario (Proposition 4.7) and the no information sharing scenario (Proposition 4.5). Consequently, the profit effect is zero, if the stockpiling segment is smaller than its critical size.

Further, if $\beta_{1t} > \tau$, we need to show that

$$\Delta\pi_k^* = \frac{\beta_2}{2}\left(\hat{f}^*\hat{\Omega}_8(r) - \check{f}_k^*\check{\Omega}_k(r)\right) > 0$$

is fulfilled. Hence it is sufficient to proof that

$$\frac{\hat{f}^*}{\check{f}^*} > \frac{\check{\Omega}_k(r)}{\hat{\Omega}_8(r)},$$

which always holds true as given by the results of Propositions 4.9 and 4.8. □

The proposition suggests that retailers only attain a positive profit effect if the size of the stockpiling segment is large enough, i.e., $\beta_{1t} > \tau$. In this case, the retailer has lower inventory costs associated with promotions (positive inventory effect) and consequently increases his promotion frequency in equilibrium (positive frequency effect). These two effects add up to a positive profit effect, making the retailer better off with information sharing. However, if the stockpiling segment is smaller, i.e., $\beta_{1t} < \tau$, information sharing is by no means harmful to the retailer; it just leaves profits unchanged between the two scenarios.

What does this imply? If it is optimal for the retailer to promote at a higher frequency on information sharing, the stockpiling customer segment will take advantage, building up additional household inventory, and will not return for a longer time for a subsequent purchase. In the weeks after a promotion, the size of the stockpiling segment will not be large enough to incentivise the retailer to promote. In these periods following the promotion, the retailer does not derive benefits from information sharing. However, given the positive profit effect at the time of the promotion, we find that information sharing is beneficial, even when information sharing is not analyzed for its immediate but for its long-term effect.

When analyzing the impact of the different customer segments on the profit effect, we find

$$\begin{aligned}\Delta\pi_k^* &= \frac{\beta_2}{2}\left(\hat{f}^*\hat{\Omega}_8(r) - \check{f}_k^*\check{\Omega}_k(r)\right)\\ &= \frac{\beta_2}{2}\left(-\alpha\Delta\Upsilon + \beta_{1t}\Lambda(p) + \beta_2\Delta\Lambda\right)\\ &\quad\times\left(\frac{\hat{\Omega}_8(r)}{\beta_{1t}\hat{\Omega}_8(p) + \beta_2\Delta\hat{\Omega}_8} - \frac{\check{\Omega}_k(r)}{\beta_{1t}\check{\Omega}_k(p) + \beta_2\Delta\check{\Omega}_k}\right),\end{aligned}$$

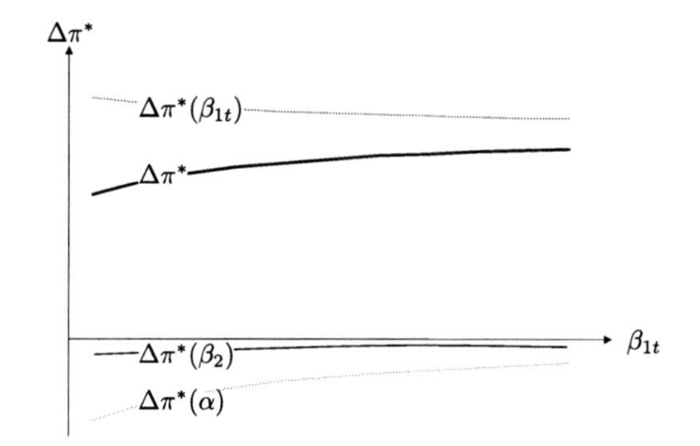

Fig. 4.13 Profit effect and the impact of customer segments

which is visualized in Fig. 4.13. As is evident, the loyal customer segment reduces the profit effect: at the promotion price, the retailer loses sales margin to the loyal customer segment who would also have purchased the product at the regular price.

Given its small size, the store-switching customer segment β_2 has a minor impact on the profit effect, however, its impact is negative for the example of Fig. 4.13. As the store-switching customers are in the market even if no promotion is offered, a promotion results in a loss of sales margin, as it does for the loyal customer segment.

Finally, the only positive, but by far largest effect, originates from the stockpiling customer segment. Given that the symmetric retailers promote at a higher frequency with information sharing than with no information sharing, they successfully exploit the stockpiling customer segment.

4.4.4 Customer Welfare

The benefit of information sharing from the customer perspective can be evaluated with regard to customer welfare. The customer would benefit if prices decrease with information sharing compared to the no information sharing scenario. As the three discrete price levels p_l, p_h and r are identical for both scenarios, the expected price paid by the customer depends solely on the promotion frequency f and the promotion depth ϕ. We shall compare the prices paid by each of the three customer segments, loyal customers α, stockpiling customers β_{1t} and store-switching customers β_2, in both scenarios.

The loyal customer segment only purchases the product at its preferred retailer. Given that ϕ is constant and exogenous for both scenarios, the expected price paid by the loyal customer segment p_α is

$$p_\alpha = f(\phi p_l + (1-\phi)p_h) + (1-f)r.$$

We denote the price paid by the loyal customer in no information sharing as $\hat{p}_\alpha$ and the one paid in the scenario of information sharing as $\check{p}_\alpha$. The impact of information sharing on the price of the loyal customer segment is summarized in the following proposition.

Proposition 4.11. *The loyal customer segment α pays a lower price in information sharing as compared to no information sharing, i.e.,* $\check{p}_\alpha \leq \hat{p}_\alpha$, $\forall \hat{f} \leq \check{f}$.

Proof.

$$\begin{aligned}
\check{p}_\alpha &\leq \hat{p}_\alpha, \\
\check{f}\big(\phi p_l + (1-\phi)p_h\big) + (1-\check{f})r &\leq \hat{f}\big(\phi p_l + (1-\phi)p_h\big) + (1-\hat{f})r, \\
(\check{f} - \hat{f})\big(r - \phi p_l - (1-\phi)p_h\big) &\geq 0.
\end{aligned}$$

This is true for $\check{f} \geq \hat{f}$, which was shown in Proposition 4.9. □

Obviously, the more frequently a retailer promotes, the lower the price for the loyal customers. But not only the loyal customer benefits from a higher promotion frequency but also the smart customer segments.

By definition the smart customer segments β_t will always pay the lowest price offered in the market:

$$p_{\beta_t} = \min[p_i, p_j] \quad \in [p_l, p_h]. \tag{4.42}$$

Further, the stockpiling customer segment β_{1t} withholds purchases if both retailers A and B offer the regular price only. Table 4.12 shows the price paid by the switching customer segment based on the offer of both retailers.

The expected price paid by the stockpiling customer segment β_{1t} is then determined by the weighted average of the minimum prices of both retailers. Observe that in case both retailers offer at regular price, the stockpiling customers do not generate any demand, which occurs with a probability of $(1-f)^2$:

$$\begin{aligned}
p_{\beta_{1t}} &= \frac{1}{1-(1-f)^2}\big(p_l f\phi\big(f\phi + 2f(1-\phi) + 2(1-f)\big) \\
&\qquad + p_h f(1-\phi)\big(f(1-\phi) + 2(1-f)\big)\big) \\
&= \frac{p_l\phi(2 - f\phi) + p_h(1-\phi)(2 - f(1+\phi))}{2-f}.
\end{aligned}$$

Table 4.12 Prices paid by the stockpiling customer segment β_{1t}

			Retailer j		
			p_l	p_h	r
			$f\phi$	$f(1-\phi)$	$1-f$
Retailer i	p_l	$f\phi$	p_l	p_l	p_l
	p_h	$f(1-\phi)$	p_l	p_h	p_h
	r	$1-f$	p_l	p_h	–

We can now summarize the impact of information sharing on this price by the following proposition, where $\hat{p}_{\beta_{1t}}$ denotes the price paid with no information sharing and $\check{p}_{\beta_{1t}}$ is the price paid with information sharing.

Proposition 4.12. *The stockpiling customer segment β_{1t} pays a lower price with information sharing as compared to no information sharing, i.e., $\check{p}_{\beta_{1t}} \leq \hat{p}_{\beta_{1t}}$, $\forall \hat{f} \leq \check{f}$.*

Proof. Comparing the expected price by the stockpiling customer segment across scenarios leads to

$$\check{p}_{\beta_{1t}} \leq \hat{p}_{\beta_{1t}},$$
$$\frac{p_l\phi(2-\check{f}\phi)+p_h(1-\phi)(2-\check{f}(1+\phi))}{2-\check{f}} \leq \frac{p_l\phi(2-\hat{f}\phi)+p_h(1-\phi)(2-\hat{f}(1+\phi))}{2-\hat{f}},$$
$$(\check{f}-\hat{f})(p_h-p_l) \geq 0,$$

which holds true if $\check{f}-\hat{f} \geq 0$ and has been shown in Proposition 4.9. □

Finally, we consider the impact of information sharing on the store-switching segment β_2. The expected price of the store-switching segment is calculated as

$$\begin{aligned} p_{\beta_2} &= p_l f\phi\big(f\phi+2f(1-\phi)+2(1-f)\big) \\ &\quad + p_h f(1-\phi)\big(f(1-\phi)+2(1-f)\big)+r(1-f)^2 \\ &= p_l f\phi(2-f\phi)+p_h f(1-\phi)\big(2-f(1+\phi)\big)+r(1-f)^2 \end{aligned}$$

and we determine the impact of information sharing in the following proposition.

Proposition 4.13. *The store-switching customer segment β_2 pays a lower expected price with information sharing than with no information sharing, i.e., $\check{p}_{\beta_2} < \hat{p}_{\beta_2}$*

Proof. We need to show that

$$\check{p}_{\beta_2} < \hat{p}_{\beta_2},$$
$$2(\check{f}-\hat{f})\big(\phi p_l+(1-\phi)p_h-r\big) < (\check{f}^2-\hat{f}^2)\big(\phi^2 p_l+(1-\phi^2)p_h-r\big),$$
$$\frac{2}{\check{f}+\hat{f}} > \frac{\phi^2 p_l+(1-\phi^2)p_h-r}{\phi p_l+(1-\phi)p_h-r}.$$

Given that $\check{f}, \hat{f} \in [0,1]$, the left hand side of the above equation is

$$\frac{2}{\check{f}+\hat{f}} \geq 1,$$

whereas the right hand side of the equation is

$$\frac{\phi^2 p_l + (1-\phi^2)p_h - r}{\phi p_l + (1-\phi)p_h - r} < 1,$$
$$(p_h - p_l)\phi(1-\phi) > 0,$$

the price paid by the stockpiling customer with information sharing is lower than the one paid with no information sharing. □

As presumed by intuition, the expected price paid by loyal as well as smart customers is lower in the scenario of information sharing. The result is linked to the fact that it is beneficial for the retailer to increase his promotion frequency with information sharing (positive frequency effect). The customers therefore benefit from the higher promotion frequency triggered by information sharing.

In Fig. 4.14, the price difference between the two scenarios $\Delta p = \hat{p} - \check{p}$ is shown for an increasing frequency effect. We see that the larger the frequency effect, the lower the prices customers have to pay with information sharing. Further, when comparing the impact of information sharing for the three customer segments, information sharing is most beneficial to the store-switching customer segment β_2, followed by the loyal customer segment α. The stockpiling customer segment β_{1t} attains the lowest benefit from information sharing, due to the low prices this segment already pays with no information sharing as a result of its bargaining mentality and willingness to empty the household inventory when no promotion is offered.

Finally, observe that the loyal and store-switching customer segments benefit from the bargaining mentality of the stockpiling customer segment: as shown in Proposition 4.9, the larger the size of the stockpiling segment, the higher the incentive for retailers to promote with information sharing and the larger the promotion frequency. A larger promotion frequency further implies decreasing prices, especially for the loyal and stockpiling customer segment.

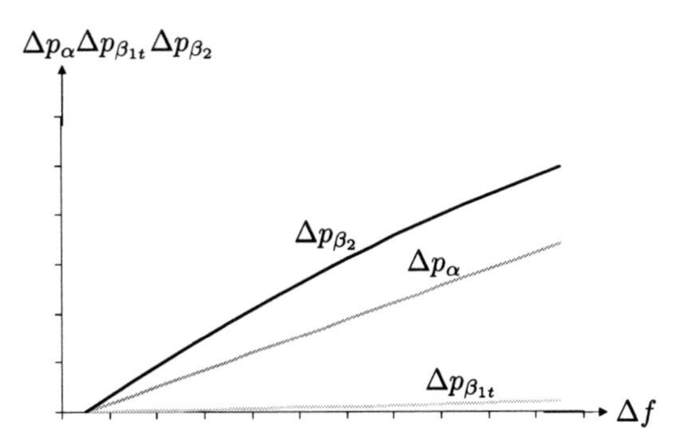

Fig. 4.14 Impact of frequency effect on customer prices

4.4.5 Manufacturer Benefits

Our focus is on the value of upstream information sharing and on deriving the benefits from a retailer's and customer's perspective. However, it is interesting that upstream information sharing also provides benefits to the manufacturer.

Private label manufacturers attained a significant market share within the last years due to the bargaining mentality of customers. Lower prices for customers, facilitate competition with private label manufacturers, and may put the branded goods manufacturer in a position, where he is able to regain market share from private label manufacturers. Especially, since the lower prices are the result of an increased supply chain efficiency and do not impose any margin effects on the manufacturer. Upstream information sharing allows to close the gap between branded goods and private labels.

Further, information sharing is legally sound since the customers do benefit. Both the loyal and the smart customer segments pay lower prices with information sharing. Upstream information sharing should thus be in the interest of the manufacturer.

And there is another aspect: in order to compose the Competition Index, the manufacturer receives early information on retailers' planned promotions. This information allows him to smooth production planning and, in addition reduce inventory costs at the manufacturer's side. As Iyer and Ye (2000) reveal, information sharing mitigates some of the costs imposed on the manufacturer by the high inventories required to support promotions. Further, the authors found that downstream information in promotions is of special importance if the stockpiling customer segment is large.

4.5 Extension: Asymmetric Retailers

In this section, we shall extend the model formulation to asymmetric retailers, in order to enable a meaningful interpretation of the data set laid out in Chap. 5. We assume that the retailers are asymmetric in their loyal customer segments and assume further that retailer A has a smaller loyal customer segment than retailer B, i.e., $\alpha_A < \alpha_B$, while keeping inventory costs and price levels constant across competitors. Further, we keep up the assumption of an exogenously set promotion depth ϕ which is the same for the two retailers. In the following, we shall first derive the mixed strategy equilibrium for asymmetric retailers and then provide the value of information sharing.

4.5.1 Asymmetric Mixed Strategy Equilibrium

We will take three steps in deriving the asymmetric mixed strategy equilibrium. First, we set up the profit function for retailers A and B. From there, we derive the best response in order to finally set up the asymmetric mixed strategy equilibrium. The following are general details which are applied to both the information sharing and no information sharing scenarios.

Profit Function

From (4.4), we can reformulate the profit function for the asymmetric retailers A and B as

$$\pi_A(f_A, f_B, \phi) = f_A \pi_A(\mathsf{p}, f_B, \phi) + (1 - f_A)\pi_A(r, f_B, \phi),$$
$$\pi_B(f_A, f_B, \phi) = f_B \pi_B(\mathsf{p}, f_B, \phi) + (1 - f_B)\pi_B(r, f_B, \phi).$$

Best Response

According to Definition 4.2, the best response is calculated as

$$\begin{aligned}\frac{\partial \pi_A(f_A, f_B, \phi)}{\partial f_A} &= \delta_A(f_B, \beta_{1t}) \\ &= \pi_A(\mathsf{p}, f_B, \phi) - \pi_A(r, f_B, \phi) \\ &= -\Delta\Upsilon_A + \Delta\Lambda_k - f_B \Delta\Omega_k, \end{aligned} \tag{4.43}$$

$$\begin{aligned}\frac{\partial \pi_B(f_A, f_B, \phi)}{\partial f_B} &= \delta_A(f_A, \beta_{1t}) \\ &= \pi_B(\mathsf{p}, f_B, \phi) - \pi_B(r, f_B, \phi) \\ &= -\Delta\Upsilon_B + \Delta\Lambda_k - f_A \Delta\Omega_k. \end{aligned}$$

The results of Proposition 4.1 hold true for the asymmetric case. That is, whenever the size of the stockpiling segment β_{1t} is smaller than its critical size τ_A, it is optimal for retailer A to refrain from promoting and rather offer the product at the regular price, with

$$\tau_A = \frac{\alpha_A \Delta\Upsilon - \beta_2 \Delta\Lambda_k}{\Lambda_k(p)}$$

and

$$\beta_{1t} < \tau_A \quad \rightarrow \quad f_A^* = 0.$$

In contrast, when the stockpiling segment β_{1t} is larger than τ_A, it is optimal for the retailer to play strategically with

$$\beta_{1t} > \tau_A \quad \rightarrow \quad f_A^* = \begin{cases} 1 & \text{if } f_B \in [0, f_B^\dagger), \\ [0, 1] & \text{if } f_B = f_B^\dagger, \\ 0 & \text{if } f_B \in (f_B^\dagger, 1], \end{cases} \tag{4.44}$$

at

$$f_B^{\dagger} = \frac{-\Upsilon_A + \Delta\Lambda_k}{\Delta\Omega_k}.$$

The results for retailer B are found by simply reversing the indices. Observe that the loyal customer segment of retailer B is larger than that of retailer A. Intuitively, retailer B attains a higher base profit from his loyal customers, and has higher opportunity cost of a promotion which increases the critical size of the stockpiling segment as compared to retailer A, i.e.,

$$\alpha_B > \alpha_A \quad \rightarrow \quad \tau_B > \tau_A.$$

With this asymmetry, we attain three different domains for the size of the stockpiling segment:

1. $\beta_{1t} \in [0, \tau_A]$
2. $\beta_{1t} \in (\tau_A, \tau_B]$
3. $\beta_{1t} \in (\tau_B, \infty)$

Within the first domain $\beta_{1t} \in [0, \tau_A]$, the size of the stockpiling segment is small and both retailer A and retailer B pursue the dominant strategy $f_A^* = f_B^* = 0$. With an increasing size of the stockpiling segment, $\beta_{1t} \in (\tau_A, \tau_B]$, retailer A's best response becomes dependent on the competitor's promotion frequency: he plays strategically as characterized in (4.44). In contrast, retailer B still makes higher profits by serving his loyal customers at the regular price than by attracting smart customers with a promotion. Consequently, he sticks to his pure strategy, "regular" with $f_B^* = 0$. Finally, in the third domain $\beta_{1t} \in (\tau_B, 1]$ both, retailer A and retailer B, play strategically by reacting to the opponents strategy. A plot of the response correspondences for the three regions is provided in Fig. 4.15.

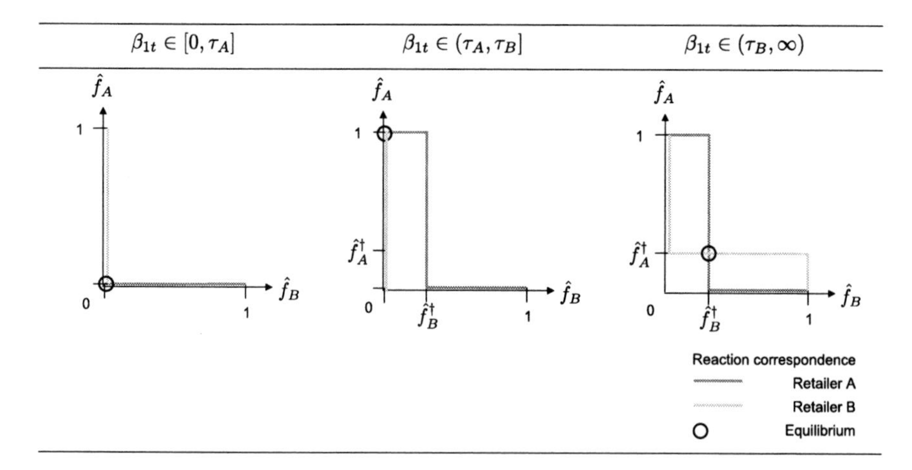

Fig. 4.15 Response correspondences for asymmetric retailers

Asymmetric Nash Equilibrium

The Nash Equilibrium is found at the intersection of the reaction correspondences in Fig. 4.15. We will summarize the asymmetric Nash equilibrium in the following proposition:

Proposition 4.14. *The Nash equilibrium for asymmetric retailers depends on the size of the stockpiling segment β_{1t}. We distinguish three cases:*

Case i: *If $\beta_{1t} \in [0, \tau]$, the game has a pure strategy equilibrium at*

$$f_A^* = f_B^* = 0.$$

Case ii: *If $\beta_{1t} \in (\tau_A, \tau_B]$, the game has a pure strategy equilibrium at*

$$f_A^* = 1,$$
$$f_B^* = 0.$$

Case iii: *If $\beta_{1t} \in (\tau_B, 1]$, the game has a mixed strategy equilibrium at*

$$f_A^* = \frac{-\Delta\Upsilon_B + \Delta\Lambda_k}{\Delta\Omega_k},$$
$$f_B^* = \frac{-\Delta\Upsilon_A + \Delta\Lambda_k}{\Delta\Omega_k},$$

with

$$\tau_i = \frac{\alpha_i \Delta\Upsilon_j - \beta_2 \Delta\Lambda_k}{\Lambda_k(p)}, \quad i \neq j = A, B.$$

Proof. Follows directly from Propositions 4.1 and 4.2. □

We show the different equilibria in the market condition matrix in Fig. 4.16. The market condition matrix visualizes the equilibrium promotion frequencies of retailer A and B for variations of the ratio of loyal customer segments $\frac{\alpha_B}{\alpha_A}$ on the y-axis subject to the size of the stockpiling customer segment β_{1t} on the x-axis. Whenever the size of the stockpiling segment is below τ_A, both retailers will offer the product at the regular price in equilibrium. In the middle section, retailer B still makes the highest profit at the regular price, in which case it is already beneficial for retailer A to offer a promotion price, and therefore succeed in gaining the entire smart customer segment. Observe that the larger the ratio of loyal customers, that is, the more loyals retailer B has compared to retailer A, the larger the section where retailer A promotes and retailer B remains at his regular price. Finally, in the third region, both retailer A and B play a mixed strategy and consequently the equilibrium is a mixed one as well. If $\frac{\alpha_B}{\alpha_A}$ reduces to 1, we obtain two symmetric retailers and, as a result, the middle section, where the strategies of the two retailers differ, becomes obsolete.

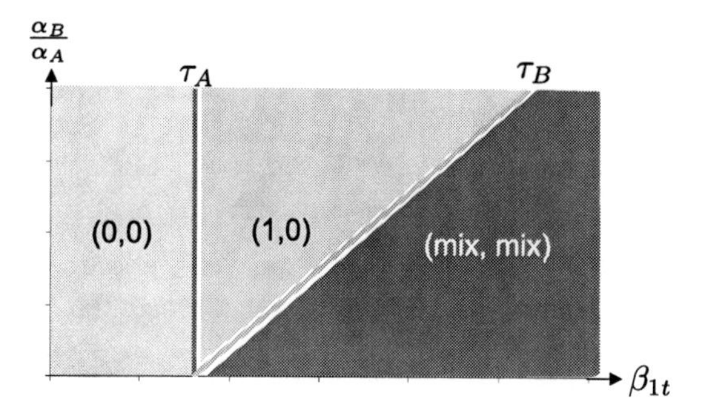

Fig. 4.16 Market condition matrix

The modified mixed strategy equilibrium under asymmetric conditions provides interesting insights into retailer strategies as also observed in the analysis of the data set (see Chap. 5): (1) A retailer's domain for choosing the pure strategy "regular", expands with an increasing share of loyal customers. Intuitively, the opportunity costs of a promotion increase with the share of loyal customers, as the reliable margin from their demand is put at stake under promotion. (2) Vice versa, the domain of the pure strategy "promote" decreases with a higher share of loyal customers. (3) Under equilibrium, the promotion frequency increases in line with the loyal customer segment, since $\alpha_B > \alpha_A$ and therefore:

$$\frac{-\Delta\Upsilon_A + \Delta\Lambda_k}{\Delta\Omega_k} > \frac{-\Delta\Upsilon_B + \Delta\Lambda_k}{\Delta\Omega_k},$$

$$f_B^* > f_A^*.$$

Consequently the critical size of the stockpiling segment τ_i and the reluctance to promote rises along with the share of loyal customers compared to competition $\frac{\alpha_B}{\alpha_A}$. On the other hand, the promotion frequency under equilibrium conditions increases, resulting in a more aggressive promotion. The explanation again goes back to the opportunity cost of putting the reliable profit margin of the loyal customer segment at stake: besides a higher entrance barrier to compete in a promotional environment, a strong retailer with a large share of loyal customers will ensure the success of his promotion by a high frequency, compensating a high margin loss with an even higher demand gain. He will therefore go all the way and play "pass or fail" if the size of the stockpiling segment is sufficiently appealing.

The equilibrium profit equations are provided in Table 4.13. Retailer B is initially passive between τ_A and τ_B, and looses the store-switching customers β_2 to retailer A. He only reacts with a successful promotion only after β_{1t} has surpassed his promotion threshold of τ_B gaining the entire stockpiling customer segment.

The market condition matrix allows us to position the retailers, and to interpret their pricing strategy in Chap. 5.

Table 4.13 Equilibrium profits for asymmetric retailers A and B

	Equilibrium	Retailer A	Retailer B
	f_A^*, f_B^*	$\pi_A(f_A^*, f_B^*)$	$\pi_B(f_A^*, f_B^*)$
$\beta_{1t} \in [0, \tau]$	$0, 0$	$\left(\alpha_A + \frac{\beta_2}{2}\right)(r - w)$	$\left(\alpha_B + \frac{\beta_2}{2}\right)(r - w)$
$\beta_{1t} \in (\tau_A, \tau_B]$	$1, 0$	$\alpha \Upsilon(p) + (\beta_{1t} + \beta_2)\Lambda_k(p)$	$\alpha_B(r - w)$
$\beta_{1t} \in (\tau_B, \infty)$	f_A^*, f_B^*	$\alpha_A \Upsilon(r) + \frac{\beta_2}{2}\Lambda_k(r) - f_B^* \frac{\beta_2}{2}\Omega_k(r)$	$\alpha_B \Upsilon(r) + \frac{\beta_2}{2}\Lambda_k(r) - f_A^* \frac{\beta_2}{2}\Omega_k(r)$

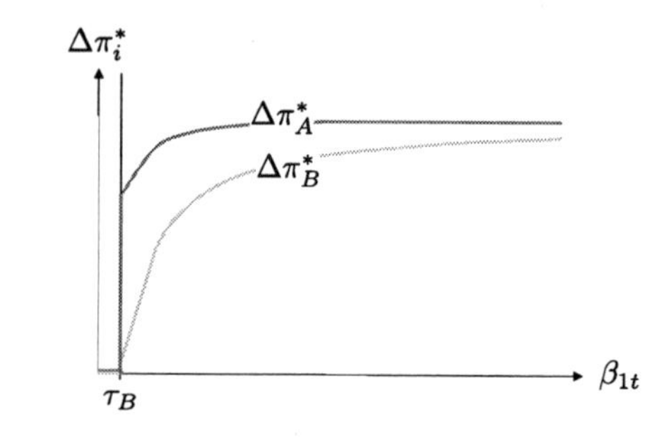

Fig. 4.17 Profit effect for retailer A and retailer B

4.5.2 The Value of Information Sharing

The results from the previous chapter on the value of information sharing hold for the asymmetric case, but require even more stockpilers in the market to come into effect, i.e., $\beta_{1t} > \tau_B$.

If the size of the stockpiling segment is smaller, the retailers will play their equilibrium strategies as described in Proposition 4.14, and given that the equilibrium strategies are known to the respective opponent, information sharing will not change the situation as compared to the no information sharing scenario. Hence, the profit effect for the retailer is zero. That is, it will neither provide additional benefit, nor worsen the profit situation of either retailer.

If the size of the stockpiling segment exceeds its critical size τ_B, the retailer will attain a positive inventory and frequency effect, as described in the previous chapter. These results are not impacted by the size of the loyal customer segment. Further, the profit effect is positive for both retailers but smaller for retailer B, the retailer with the larger loyal customer segment.

The profit effect for asymmetric retailers A and B is visualized in Fig. 4.17, making it evident that information sharing results in a win–win situation for either retailer, if the size of the stockpiling segment is large enough. For smaller sizes of the stockpiling segment, the profit effect is zero.

Further, customer welfare increases under information sharing with asymmetric retailers, and moreover as compared to symmetric retailers: in the asymmetric case, retailer B, the larger retailer, promotes more frequently in the asymmetric equilibrium than in the symmetric equilibrium. Consequently, the benefits for the individual customer segments are larger.

We shall provide a numerical study on the magnitude of impacts in the Chap. 5.

4.6 Summary

We conclude this chapter with a short comparison of the impact of information sharing on retailer profits and the welfare of different customer segments in the symmetric and asymmetric case.

Table 4.14 Comparison of the impact of information sharing

	$\beta_{1t} < \tau_B$	$\beta_{1t} > \tau_B$		
	Symmetric and asymmetric	Symmetric	Asymmetric	
			Retailer A	Retailer B
Retailer profit $\Delta\pi$	0	++	+++	+
Loyal customer α_i	0	+	+	+
Stockpiling customer β_{1t}	0	+	+	
Store switching customer β_2	0	+	+	
Manufacturer	0	+	+	

Legend: Benefit from information sharing
0 none
+ positive

From Table 4.14, it becomes evident that sharing the information from the Competition index results in a joint win–win–win situation for retailers and customers as well as for the manufacturer. Neither retailers nor customers will lose under information sharing. However, the magnitude of effect depends on the size of the stockpiling customer segment.

With a large stockpiling customer base, retailers can increase their profits in either the symmetric and the asymmetric case, with the profit increase being highest for the retailer with the smaller share of loyals, i.e., retailer A. In order to win the profits from the smart customer segment, retailer A is willing to put the profits from his smaller loyal customer base at stake. This makes him highly competitive.

Both the symmetric and the asymmetric retailers profit benefits from an improved match of supply and demand under information sharing. The reduced inventory cost makes it optimal to promote more frequently, which in turn is positive for all three customer segments: loyals, stockpilers and store-switchers pay lower prices in the scenario of information sharing.

Moreover, we have qualitatively shown that the manufacturer benefits from upstream information sharing. The lower prices paid by the customers allow the branded goods manufacturer to close the gap between private labels and branded goods, making him more competitive in the market place.

Chapter 5
Empirical Analysis

We support the findings of our analytical study with empirical evidence from the diapers category. We begin the analysis by describing our data set and the different retail chains that form the competitive landscape, in the first section. In the second section, we give a brief profile of each retail chain before analyzing their pricing levels with respect to regular and promotion prices, as well as their promotion strategies, including retail price format and frequency.

Further, we provide an in-depth analysis of customer demand, depending on the prices and quantities offered by the retailers, in order to understand the price sensitivity of customers with the aim of decomposing customer demand into segments of loyal and smart customers. We further distinguish smart customers into store-switching and stockpiling customers according to the segmentation in Sect. 4.1.3. From there, we derive the segmentation-based retailer strategies that can be observed in the competitive landscape.

Subsequently, we include a second data set containing order quantities of the retail chains on the supply side in order to analyze the correlation between order quantities and the resulting competitive pressure for future periods. We compile the Competition Index and show how this piece of information can be valuable for the retailer in order to forecast promotion activities of his competition in upcoming periods.

We conclude the chapter with a numerical study on the value of information sharing for a symmetric retailer and the different customer segments.

5.1 The Data Set

In this section, we describe the data set applied for the analysis as well as the different retail chains that form the competitive landscape, and the relevant players on the retailer side, including their package sizes, product offer and brand focus.

D. Wiehenbrauk, *Collaborative Promotions*, Lecture Notes in Economics and Mathematical Systems 643, DOI 10.1007/978-3-642-13393-0_5,

5.1.1 Description of the Data Set

The empirical analysis is based on market data from the diapers category. We chose this category for three reasons. First, brand switching is uncommon with diapers because, typically, customers initially decide for one brand, and are brand loyal for the remaining usage time. Second, diapers are rather expensive and budget-conscious parents are price sensitive and behave in a smart way in the market (Huchzermeier et al. 2002). That is, they react to promotions with store-switching and stockpiling in order to get a good deal. Third, stockpiling does not induce the "consumption effect" because the usage rate of diapers is fairly constant.

The category provides the advantage of focusing on those customers that switch between stores and thereby contribute to the high demand uncertainty faced by the retailers. The objective of sharing the upstream information in the form of the Competition Index is to reduce the demand uncertainty and hence the forecast error due to stockpiling and store-switching behavior. The characteristics of the data set enable a controlled and focused study.

Our data set comprises two different sources, one supply side and one demand side as shown in Fig. 5.1. First, we obtained delivery quantities from the manufacturer aggregated across retailers in the German market. The data set contains monthly data for the period from July 2002 until December 2004. The aggregated delivery quantities allow us to simulate the Competition Index. The second data set contains information about the aggregated diapers sales for 121 weeks (week 36/2002 to week 52/2004) of six major German retail chains denoted as retailers A–F. For each of these chains, the data set contains sales volumes and

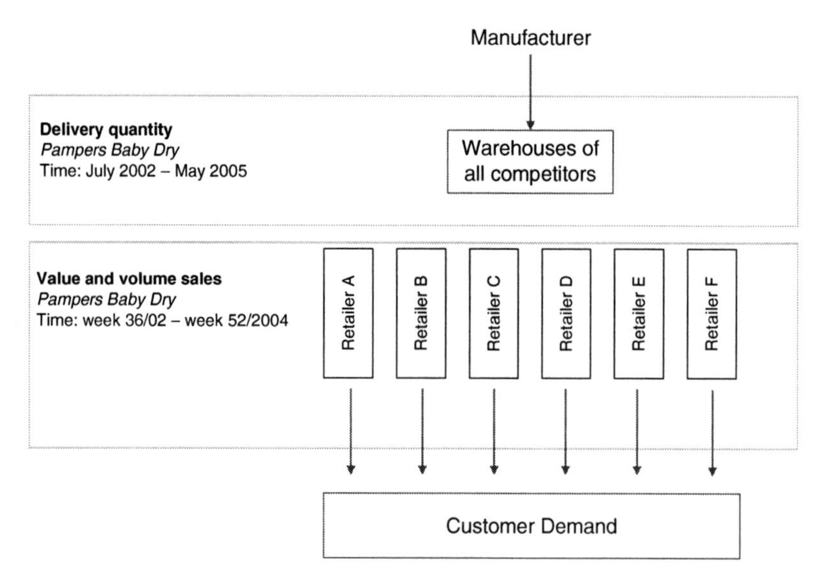

Fig. 5.1 Description of data set

sales values on a weekly basis for the diapers brands of Pampers, Fixies and Private Labels, as well as the share of promoted sales volume. The empirical estimation of effects in the following sections focuses on the full 2-year period from January 2003 to December 2004.

5.1.2 Competitive Landscape

The six major German retail chains form the basic population of retailers in our analysis. Together they achieve a category market share of 93% with respect to diapers and 94% for Pampers in particular and can therefore be considered as representative for the entire category. Hard discounters are not considered since they did not sell premium brands in the observed period at all. We will analyze the characteristics of these retail chains along the dimensions of (1) store format, (2) category market share, (3) Pampers market share, (4) brand share and (5) package size. The profiles of the retail chains are summarized in Fig. 5.2.

The three relevant store formats are hypermarkets, supermarkets and drugstores, which are defined below:

A hypermarket is a retail store with a minimum selling space of 5,000 m^2 and about 33,000–63,000 different products which are mainly offered in the self-service mode. In addition to food, the range of products comprises durables and consumables. Hypermarkets are typically located outside city centers (Metro Group 2006).

Supermarkets are retail stores with a selling space of 400 m^2 or more, offering food and luxury items, including fresh produce and complementary short-term necessaries or other lines, mainly in the self-service mode. The assortment of a supermarket comprises about 7,000–12,000 items; the space for nonfood is usually limited to 25% of overall selling space. If the selling space covers 1,500 m^2 or more, the stores are referred to as superstores, and from 5,000 m^2 as hypermarkets (Metro Group 2006).

Drugstores specialize in sales of personal hygiene, cosmetics, health food, cleaning agent, detergent and other household-related nonfood items. The selling space of drugstores varies from a relatively small 100 m^2 to more than 1,000 m^2 with an assortment ranging from 4,000 to 25,000 items (Metro 2006). They are typically located within residential areas and pedestrian zones often without car access, and therefore focus on easy-to-carry and smaller sized packaging.

While retailer A is a pure hypermarket chain, retailers B, D and E operate supermarkets as well as larger hypermarkets. Retailers C and F can be classified as drugstores. The number of sales points is characteristic for the store format. While pure hypermarket chains focus on major urban areas with as few as 200–300 stores, the large cross-regional supermarket chains usually operate 2,000–4,000 stores across Germany. A few drugstore chains even exceed 10,000 outlets with a predominant presence mainly in pedestrian areas.

Retailer	Store Format	Category Market Share	Pampers Market Share	Brand Share	Package Size Share
Retailer A	Hyper-market	15%	22%	10%, 16%, 73%	5%, 2%, 93%
Retailer B	Super-/Hyper-market	13%	17%	15%, 15%, 70%	32%, 32%, 35%
Retailer C	Drug store	27%	17%	42%, 32%, 26%	100%
Retailer D	Super-/Hyper-market	13%	16%	17%, 17%, 65%	47%, 34%, 20%
Retailer E	Super-/Hyper-market	4%	5%	18%, 14%, 68%	30%, 36%, 34%
Retailer F	Drug store	21%	17%	53%, 42%, 5%	15%, 85%
Sum of Retailers A-F	**n/a**	**93%**	**94%**	31%, 52%, 17%	16%, 47%, 37%

Pampers / Fixies / Private Label
Value Pack / Jumbo Pack / Mega Pack

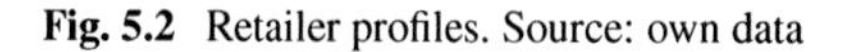

Fig. 5.2 Retailer profiles. Source: own data

The diaper category market share of supermarkets and hypermarkets adds up to 45% in Germany, while drugstores cover 48%. The remaining 7% can be assigned to individual retail stores, gas stations, etc. While the drugstores head the overall category market share, supermarkets and hypermarkets dominate the sales of the premium brand, Pampers, with retailer A as the market leader, 5% ahead of Retailers B and C. The Pampers market shares show some variation among the retail chains

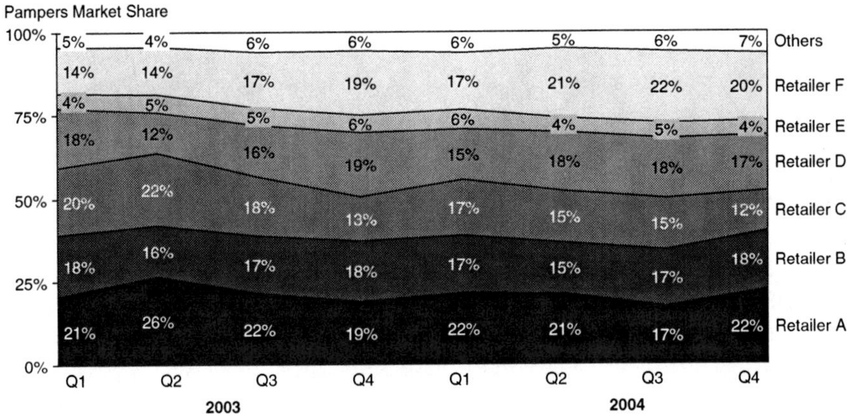

Fig. 5.3 Pampers market share by retailer. Source: own data

in the period considered from 2003 to 2004. Although with a slight decrease, the four market leaders, i.e., retailer A, B, C and D, account for more than 70% of sales across the period as shown in Fig. 5.3.

The brand share describes the share of the two premium brands, Pampers and Fixies, versus private labels at each retailer. In total, Pampers is the market leader with 52% of total sales, followed by Fixies, with 17%. The aggregated private labels make up the remaining 31% of diaper sales. With regard to store formats, supermarkets and hypermarkets all sell more than 80% of the premium products, Pampers and Fixies, and only a low share of private labels. The opposite is true for the drugstores, where private labels are predominant with 42% and 53% respectively.

The package size is considered for the brand Pampers only. With regard to package size, drugstores predominantly sell small package sizes such as Value Packs (40–52 units depending on the diaper size), while hypermarkets and supermarkets tend to larger package sizes such as Jumbo (80–104 units) or Mega Packs (160–208 units), driven by the shopping behavior of the store formats: while shopping at hypermarkets and supermarkets takes place on a more regular basis with stockpiling intentional, at least in the short-term, drugstores are frequently used as an ad hoc source within walking distance.

5.2 Retailer Strategies

After having characterized the market players in Germany's retail environment in general and the diapers market in particular, we proceed with a description of their different approaches in competing for customers in order to validate the assumptions on retailer pricing strategies described in Sect. 4.1.2.

First, we analyze the price ranges at which they offer the product in focus: diapers of the brand, Pampers Baby Dry. We continue by identifying the price levels at which each retailer offers the product to the customers under promotion and non-promotion conditions, and further describe the promotion pattern and frequency at which the retailers promote. We conclude with a brief characterization of each retail chain's strategy, taking into account the previously specified profiles.

5.2.1 Price Ranges

The prices offered by the retail chains and individual stores vary weekly due to the promotion decisions taken by the retailers. Figure 5.4 illustrates the ranges of average weekly prices for Pampers Baby Dry for all retailers including average, standard deviation and quartiles in a box-and-whisker plot. The average weekly price is calculated by dividing the national weekly sales value in EUR of each retail chain for the given product by the respective sales volume in units. In order to find a common denominator for all retail chains and account for different package sizes, we uniformly base our further analysis on statistical units (SU) that in turns consist of 180 units (diapers) each.

Differences in average weekly prices are caused by several parameters that are subject to the retail chain's price levels and promotion strategy, comprising both retail price format and promotion frequency. While for all retailers but retailer C the average price is between 0.21 and 0.22 EUR/unit, the average price of retailer C is at a high of 0.25 EUR/unit. This is mainly caused by the fact that retailer C sells exclusively small package sizes as evident from Fig. 5.3. The observation that the unit price varies with the package size has also been made in (Huchzermeier et al. (2002)). In addition, retailer C shows the largest price spread among the competitors,

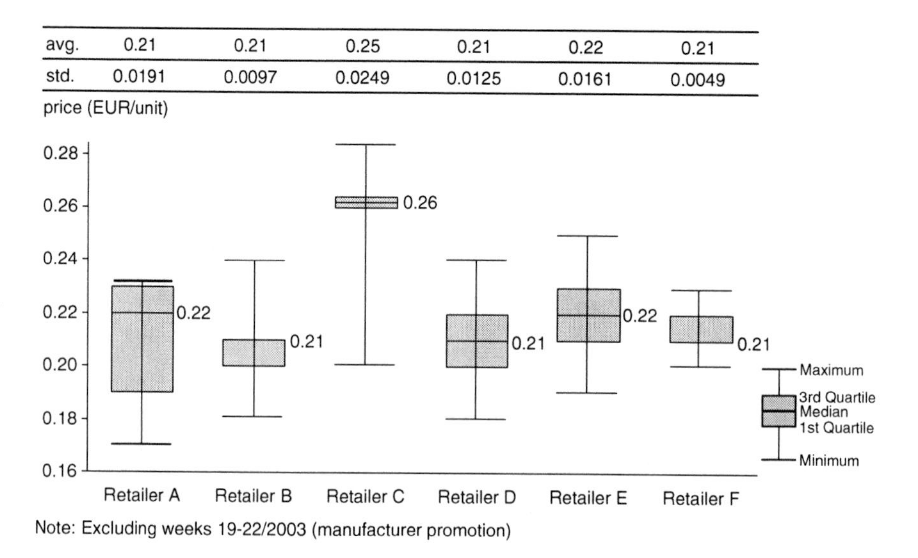

Fig. 5.4 Price ranges and standard deviation by retailer (EUR/unit). Source: own data

leading also to the highest standard deviation of 0.0249. The narrow band between the first and third quartile of retailer C, combined with a large standard deviation, implies a less frequent but therefore more intense promotion scheme. On the other hand, the large spread between the first and third quartile of retailer A, compared to the other retailers, suggests a high use of the available pricing range with frequent pricing at the upper and lower bounds. Retailer F operates within the smallest pricing range of all competitors at a low level and very low standard deviation, which suggests an everyday low pricing strategy (EDLP).

The following week-by-week analysis of average prices offered by each retailer will give further insights on each retailer's price levels under promotion and regular sales and sets the stage for the analysis of promotion patterns and frequency.

5.2.2 *Price Levels*

Given that most retail chains offer products at regular prices and promotion prices based on their promotion strategy, we observe three price levels at each retailer (Figs. 5.5 and 5.6). We distinguish between a regular price r, a high promotion price p_h and a low promotion price p_l in analogy to the model setup in Sect. 4.1.2. The retailer offers the product at a regular price r in a non-promotion period. The regular price is the upper price limit approaching the loyal customers' reservation price and is usually delimited by the Manufacturer Suggested Retail Price (MSRP). It may, however, be adjusted by retailers over time to reflect the competitive situation, e.g., due to inflation, the introduction of new products or an adjustment of the general pricing strategy. The difference of regular prices among retailers originates, to a large extent, from the predominant package size offered at the retailer, where smaller package sizes as in the case of retailer C go along with a higher regular price. Further, the average weekly prices deviate from the discrete price levels as a result of the promotion frequency being continuous.

In promotion periods the product is usually offered at a high promotion price p_h, which was observed to be 4 cents below the reservation price for most retailers, except retailers C and F. Less frequently, retailers decrease prices even further to the level of a low promotion price p_l, at 6 cents below the regular price, e.g., for special campaigns with more aggressive pricing in order to gain market share or to push volume temporarily. The development of average weekly prices for all retailers is shown in Figs. 5.5 and 5.6. The prices are rounded to the second digit for clarity in visualization.

For retailers A, B, D and E, we visually observe the three defined price levels r, p_l and p_h, with a constant offset of 4, or, as the case may be, 6 cents and a stepwise decrease in the considered period due to external influences. Retailer C offers promotions only at one promotion price level $p_l = p_h$, whereas for retailer F only one discrete price level can be observed from the fourth quarter of 2003 onwards and therefore $r = p_l = p_h$. The price levels for all retailers are summarized in Table 5.1. Note that the price levels vary over time as a result of price level adjustments as described earlier.

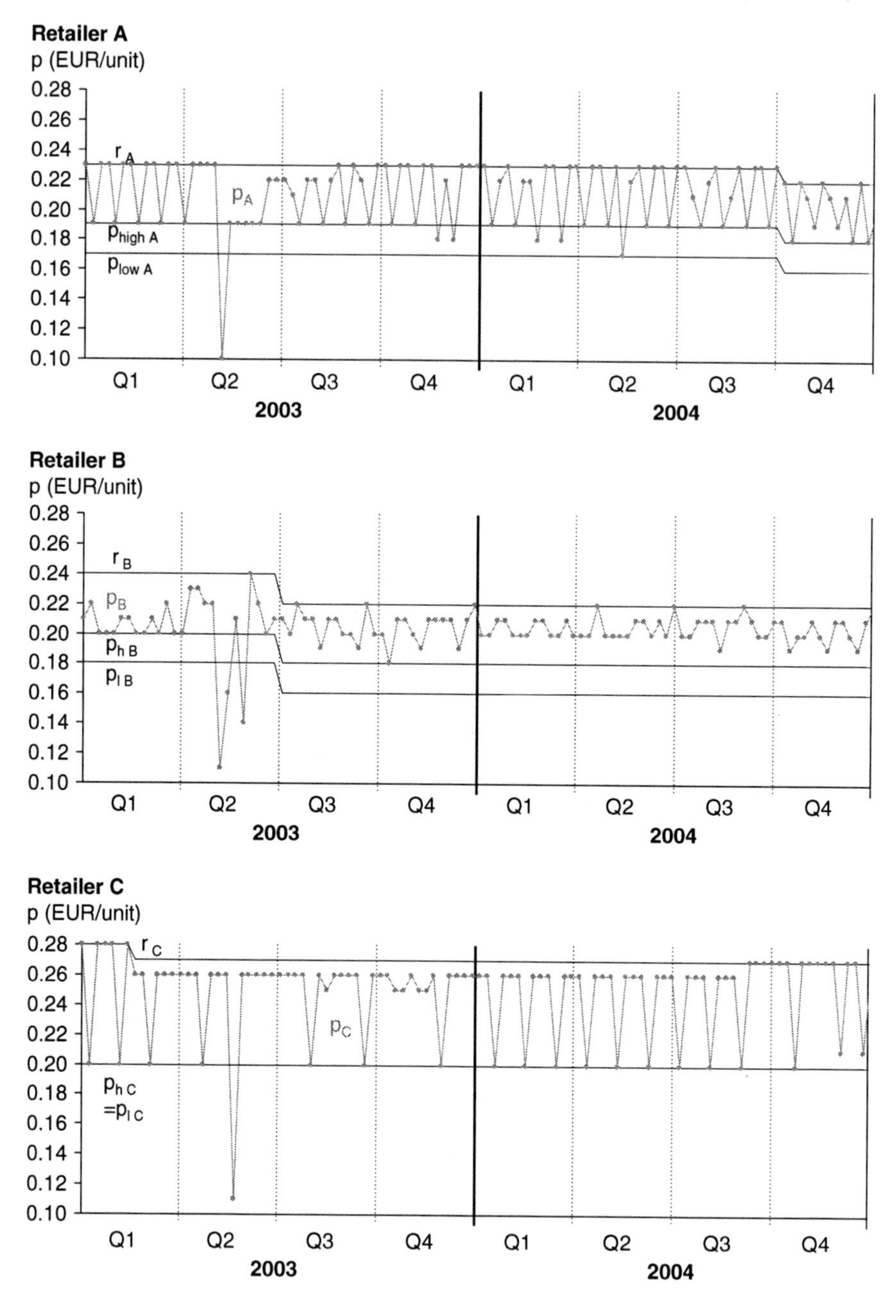

Fig. 5.5 Weekly prices for retailers A, B and C. Source: own data

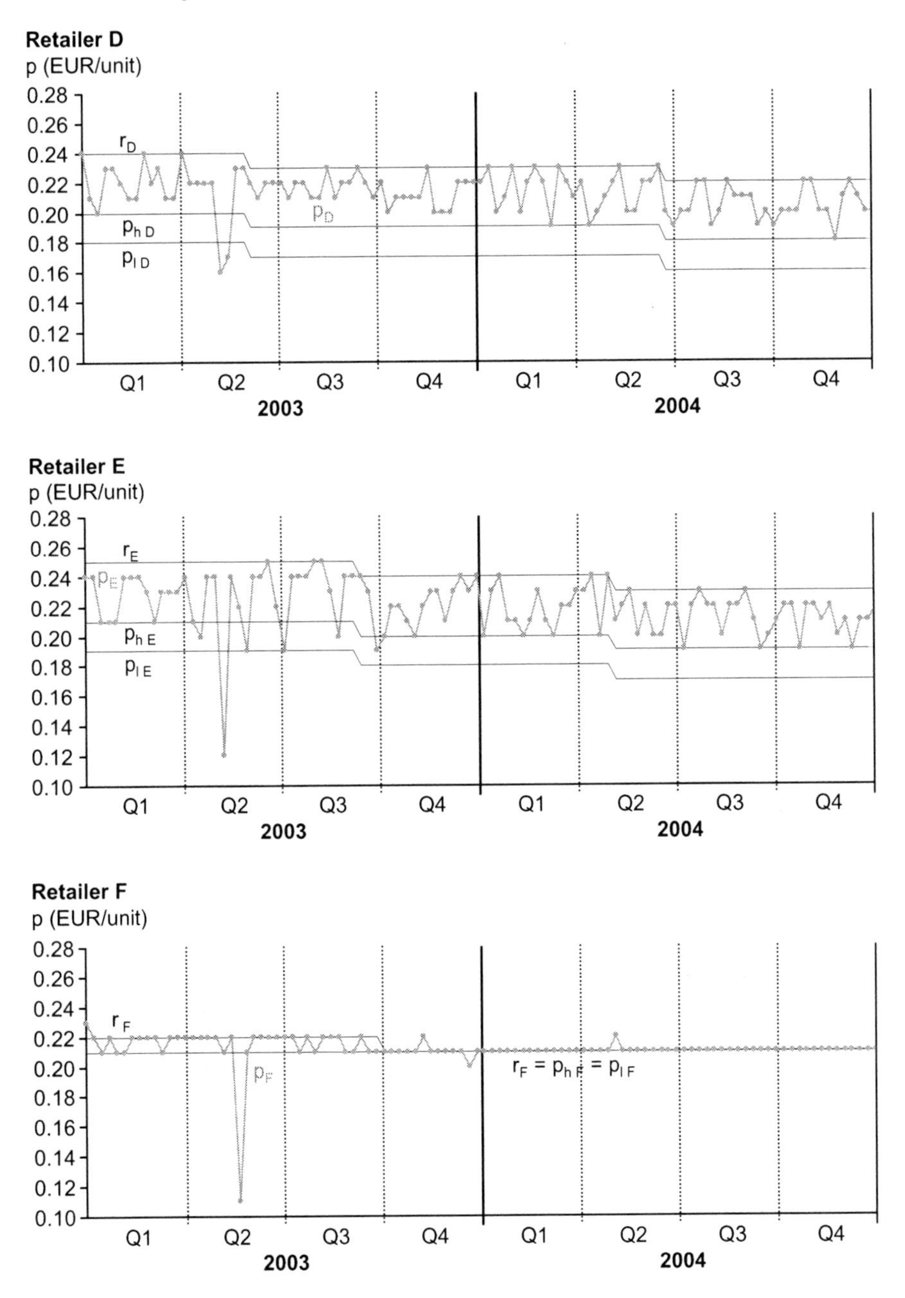

Fig. 5.6 Weekly prices for retailers D, E and F. Source: own data

Table 5.1 Range of price levels within observed period by retailer. Source: own data

EUR/unit	Average price p_{avg}	Regular price r	High promotion price p_h	Low promotion price p_l
Retailer A	0.21	0.22–0.23	0.18–0.19	0.16–0.17
Retailer B	0.21	0.22–0.24	0.18–0.20	0.16–0.18
Retailer C	0.25	0.27–0.28	0.20	0.20
Retailer D	0.21	0.22–0.24	0.18–0.20	0.16–0.18
Retailer E	0.22	0.23–0.25	0.19–0.21	0.17–0.19
Retailer F	0.21	0.21–0.22	0.21	0.21

A discount of more than 50% that can be observed for all retailers in the second quarter of 2003 was initiated by the manufacturer and therefore present in the whole market. In the given data set, the introduction of a new marketing campaign "Pampers surprises" ("Pampers überrascht") in weeks 19–22, 2003 was a one-time effect that affected the market for the following periods. These one-time effects are excluded from further analysis as the remarkably low prices and the resulting high demand were triggered by exceptional manufacturer-driven marketing activities that distorted the competitive environment among the competing retailers. Further, all retailers were facing major stockouts in the weeks of the promotion and the weeks thereafter, such that the sales volume by no means reflects actual customer demand.

5.2.3 Retail Price Format

The sequence of pricing at the different price levels can be characterized by the promotion pattern that describes the retail price format. We distinguish between different retail formats with respect to patterns as discussed in Sect. 2.2.2, namely: (1) HILO, (2) pattern and (3) Every Day Low Pricing (EDLP).

As presumed from the price ranges, the week-by-week analysis supports the assumption that Retailer A frequently offers promotions at a high discount: we find a markdown of at least 4 cents in a 3-week pattern (r, r, p-policy).

For retailer C we observe a similar regular pattern of mostly 4 weeks with a higher markdown of 6 cents (r, r, r, p-policy).

Retailer F, operating only on one price level as of the third quarter of 2003, shows very low variation in prices at a constantly low level of 0.21 EUR/unit and can therefore clearly be classified as EDLP.

Retailers B, D and E show less consistency with respect to the observed price levels as well as the duration of their promotion cycles. All three retailers apply HILO pricing. Retailer B exploits only part of the price range between the regular and the promotion price levels with a spread of 2 cents. Retailers D and E make use of the full 4 cent price range and do so in an irregular sequence, which appears to be random with respect to average weekly prices at first sight.

5.2.4 Promotion Frequency

The promotion frequency specifies the scope with which retail chains implement promotions in the market. It is denoted by f_i and specifies the share of individual stores within a retail chain that sell at a promotion price, reflecting the probability that a customer runs across a promoting store of the retail chain in a given period (compare Sect. 4.1.2). The organizational structure of a retail chain and especially the degree of centralization has a major influence on the promotion frequency. That is, if the retail chain's promotion strategy is mapped centrally and rolled out nationally, a homogeneous pricing across all stores on a weekly basis can be expected. In terms of promotion frequency, a pure centralized approach with all stores pursuing the same strategy results in two discrete frequency values of either 1 or 0, for promotion and non-promotion periods, and average prices in line with the discrete price levels, as observed for retailers A and C.

The opposite is true for retail chains with decentralized promotion decisions, where individual stores led by entrepreneurs decide on their participation in promotion activities on a weekly basis. This leads to a scattered distribution of the promotion frequency and a low adherence of the average weekly prices to the price levels as observed for retailers B, D and E. The initially observed randomness of average price levels therefore results from a heterogeneous promotion strategy within the retail chains.

In order to derive the promotion frequency from the given data set on a weekly basis we use a sales volume-based approximation for each retail chain i and week t. The promotion frequency is thereby approximated by the share of sales volume promoted at either promotion price level $d_{it}(p_l)$ and $d_{it}(p_h)$ compared to the total sales volume d_{it} in the given week. Based on the granularity of our retail environment with 250 to more than 10,000 stores per retail chain this approach delivers a valid approximation of the promotion frequency being defined as the share of stores promoting in the given week. Hence, the promotion frequency f_{it} at retailer i in week t is approximated by

$$f_{it} = \frac{d_{it}(p_l) + d_{it}(p_h)}{d_{it}}.$$

The distribution of promotion frequencies at each of the six retailers are summarized in Fig. 5.7. The distribution of the promotion frequencies confirms the retail formats of each retailer and gives further information about the degree of centralization of promotion strategies. Retailers A, C and F are characterized by a high density of promotion frequency at either bound of the frequency range. While retailers A and C frequently operate at both bounds $f = 0$ and $f = 1$, in line with their regular promotion patterns, retailer F shows a high concentration at the frequency range <0.2, which is also reflected by a low average frequency of 0.15 and characterizing EDLP behavior.

In order to reflect the two promotion price levels p_h and p_l, we will further apply a two-stage approach in deriving the promotion depth. While f_i denotes the

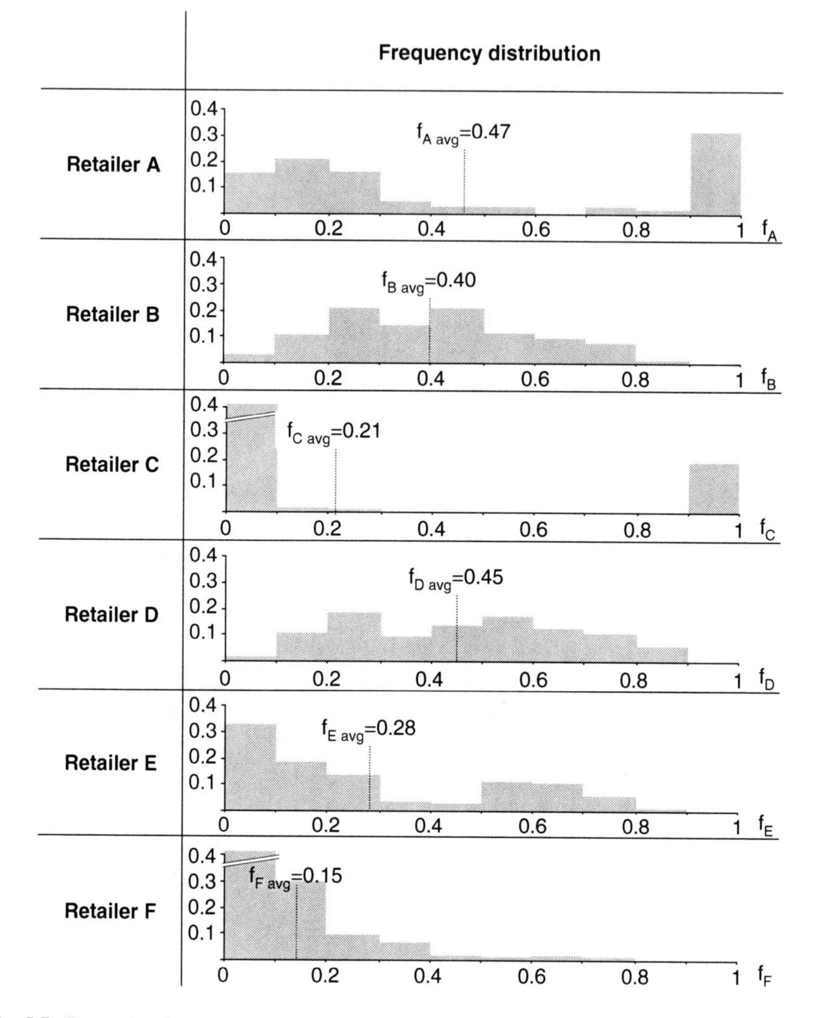

Fig. 5.7 Promotion frequency ranges and distribution by retailer. Source: own data

penetration with respect to the initial promotion decision of all stores within the retail chain, the promotion depth regarding the promotion price level is anticipated by ϕ_i as the share of promoting stores promoting at p_l. As the considered data set does not yield information on store level nor on the different promotion levels and depth, we derive ϕ_i given the promotion frequency and the defined price levels as

$$p_i = f_i\big(\phi p_l + (1-\phi)p_h\big) + (1-f_i)r \quad \rightarrow \quad \phi_i = \frac{p_i + f_i(r - p_l) - r}{f_i(p_h - p_l)},$$

where p_i is the average price across stores of retailer i as shown in Figs. 5.5 and 5.6.

Table 5.2 Promotion depth by retailer. Source: own data

	ϕ_{min}	ϕ_{avg}	ϕ_{max}
Retailer A	0	0.12	1
Retailer B	0	0.40	1
Retailer C	0	0	0
Retailer D	0	0.20	1
Retailer E	0	0.75	1
Retailer F	0	0	0

The promotion depth ϕ_i as summarized in Table 5.2 further supports the store and retail price formats of the retail chains and leads to the following observations: (1) Retailer A offers only few low promotions as evident from Fig. 5.5 but at a high depth, meaning the majority of his stores participate. (2) Retailer C's promotion decisions are perfectly centralized, offering only one promotion level $p_h = p_l$ and therefore $\phi_C = 0$. (3) Retailer F offers no promotions at all as he follows an EDLP strategy as concluded earlier. We observe a low average promotion frequency as well as a promotion depth of $\phi_F = 0$. (4) Retailers B and D make more or less frequent use of low promotions given their decentralized promotion decisions. With their promotion frequencies being more evenly spread across the frequency range as shown in Fig. 5.7, the deviation of the average weekly prices from the discrete price levels is obvious. (5) Retailer E shows a lower average promotion frequency than the other HILO retailers (B and D) with a concentration at the lower bound. In contrast his average promotion depth is at a high level with $\phi_{E avg} = 0.75$. The explanation is that a lower share of stores within the retail chain is promoting at all, while those promoting fully exploit the price between r and p_l. Again this promotion strategy leads to a high deflection from the discrete price levels in Fig. 5.6.

5.2.5 *Strategies of Analyzed Retail Chains*

Having evaluated the strategies of the six retail chains in terms of price range, price levels, retail price format and promotion frequency, we summarize the findings with a brief characterization of each retailer taking into account their profiles from Fig. 5.2.

Retailer A is a pure hypermarket chain and market leader for Pampers sales with a market share of 22%. The majority of sales (93%) are generated from large package sizes. Promotion decisions are taken in a highly centralized manner and follow a regular 3-week pattern (r, r, p-policy) with a spread of 4 cents and few lower promotions.

Retailer B, as a decentralized supermarket and hypermarket chain, offers all package sizes in equal shares. The HILO retail format, along with a wide spread in promotion frequencies, results in a low adherence to price levels on a weekly national average, and therefore a low exploitation of the price range on a national level, while individual stores might well make use of the entire range.

Retailer C, a large drug discounter chain with more than 10,000 relatively small stores, exclusively sells small package sizes. A perfectly centralized promotion strategy characterized by only two discrete promotion frequencies 0 and 1 can be observed in the regular 4-week pattern (r, r, r, p-policy) with a large spread of 6 cents between the regular and the (one) promotion price level.

Retailers D and E are similar-sized supermarket and hypermarket retail chains operated as franchises. They follow a decentralized HILO promotion strategy which results in a scattered distribution of promotion frequency and irregular average weekly prices. The individual stores of retailer E make use of a wider price range and frequently price at a low promotion price indicated by a high promotion depth.

Retailer F is the only EDLP player in the market offering a high share of small package sizes at a single low average price even for the premium brand, Pampers. Besides a high degree of private label sales, the stringent EDLP retail format demands a strict central price policy for premium brand offers as well.

5.3 Customer Demand

In the following section, we will analyze the customer demand that retailers face in the market. We begin by examining the sensitivity of the entire customer base to different prices and the price gap to the average competitor price for a retail chain.

We continue by decomposing the customer base into the customer segments (1) loyal customers α, (2) stockpiling customers β_{1t} and (3) store-switching customers β_2.

Finally, we apply the segmentation of each retailer's customer base on the asymmetric mixed strategy equilibrium from Sect. 4.5.1 to derive the segmentation based retailer strategies.

5.3.1 Price Sensitivity of Customer Demand

Based on the given data set, we analyze customer sensitivity to the average weekly price offered at retailer A. The average weekly unit price is plotted in the upper graph against the left axis in EUR/unit in Fig. 5.8. The development of weekly demand of retailer A is depicted in million units/week in the lower graph against the right axis.

The opposing peaks of price and demand indicate a high negative correlation between demand and price with an R^2 of 0.85. Although demand and price are highly negatively correlated across periods, we can, nevertheless, observe a considerable variation of demand even among periods with the same average weekly price. For example, the price offered in the two promotion periods labeled 1 and 2 in Fig. 5.8 is 0.19 EUR/unit in both cases, while the demand varies between 5.0 and 9.2 million units/week. Even in non-promotion periods with a regular price

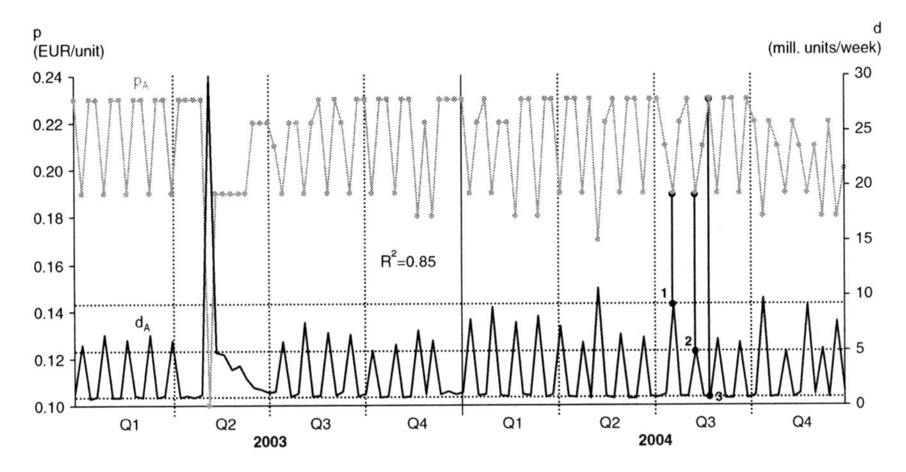

Fig. 5.8 Weekly development of price and demand at retailer A. Source: own data

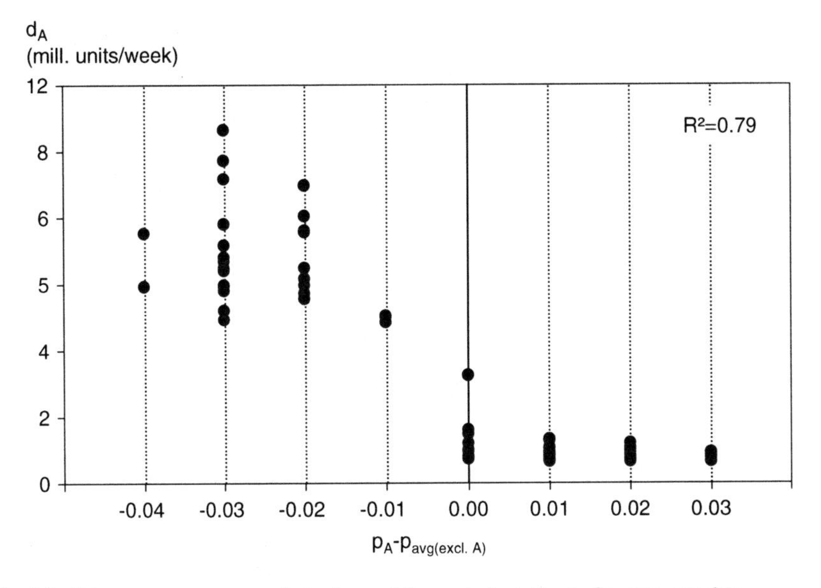

Fig. 5.9 Price gap to average market price and demand of retailer A. Source: own data

of 0.23 EUR/unit offered by retailer A, e.g., in the week labeled 3, demand varies between 0.6 and 1.0 million units/week.

We find the explanation for the variation of demand at a given price level of retailer A, when taking into account the pricing of the competitors. The demand at retailer A is subject to the price gap between the average weekly price of retailer A and the average competition price (excluding retailer A) as plotted in Fig. 5.9. The distribution of demand clearly indicates the effect the competitive price has on retailer A's demand with an R^2 of 0.79. A positive price difference implies that the price offered by retailer A is above the average competition price, leaving him with a demand in the range of 0.6–1.6 million units/week. Pricing at market average increases demand only slightly up to 2.0 million units/week, while a further reduction of the average price boosts demand to as much as 10 million units/week.

The data analysis confirms intuition that (1) customers obviously react in a highly sensitive way to decreasing prices, with an increase in demand and (2) demand is subject to the relative price positioning against competition. Note that the demand does not drop below a certain threshold, even at prices clearly above market average. A residuum of customers remains loyal to retailer A even at a higher price level: the loyal customer segment. We will further decompose the customer base in the following chapter.

5.3.2 Decomposition of Customer Demand

In order to further explain the demand variation at individual retailers in a competitive market, we continue by numerically decomposing the heterogeneous customer base into more uniform customer segments. The decomposed customer demand provides the figures for our numerical analysis of the value of information sharing in Sect. 5.5.

We therefore employ the setup derived in Sect. 4.1.3 including loyal and smart customers. We begin by determining the amount of loyal customers for each retail chain. In a second step, we identify the store-switching customer segment as being reasonable constant across periods. For modeling purposes, the size of the loyal and the store-switching customer segments are estimated on a quarterly basis. The remainder of the demand is attributed to the stockpiling customer segment with a high variation across periods on a weekly basis.

Loyal Customer Segment

As defined in Sect. 4.1.3, loyal customers α represent the least sensitive customer segment. They purchase their required quantity every week (for modeling purposes 1 unit/week) unaffected by other retailers' promotions as long as the price at their preferred retailer does not exceed their reservation price. Loyal customers are clearly assigned to their preferred retailer and stay fairly constant in the short-term with only little variation over time. In line with Blattberg and Neslin (1990), the demand from the loyal customer segment can be interpreted as the baseline demand in periods where there is no promotion.

With regard to the data set considered, the loyal customer segment is decomposed from the entire demand as the minimum demand of each retail chain within a given period. In order to respect the changing market conditions, we estimate the loyal customer segment per retail chain on a quarterly basis (13 weeks), with each

Table 5.3 Loyal customer segments by retailer. Source: own data

Units	α_A	α_B	α_C	α_D	α_E	α_F
Q1 2003	669,780	1,004,400	726,480	1,158,480	300,960	1,264,320
Q2 2003	722,160	908,820	759,060	905,580	305,640	1,308,600
Q3 2003	817,740	985,320	796,680	1,101,960	298,080	1,330,380
Q4 2003	691,200	1,175,940	792,720	1,254,960	450,000	1,712,700
Q1 2004	766,440	1,211,760	742,680	1,033,560	333,540	1,839,240
Q2 2004	634,860	1,015,380	733,860	1,204,920	320,760	1,976,760
Q3 2004	634,500	1,342,620	641,700	1,262,520	302,580	2,215,800
Q4 2004	696,420	1,251,000	556,740	1,058,040	380,700	2,200,680
Average	704,138	1,111,905	718,740	1,122,503	336,533	1,731,060
α_{iavg} / d_{iavg} (%)	23	50	36	51	51	74

Table 5.4 Average quarterly decomposition of customer segment. Source: own data

Units	$\sum \alpha_i$	β_{1t}	β_2	$\sum d_i$
Q1 2003	5,124,420	4,359,738	1,270,800	10,754,958
Q2 2003	4,909,860	2,570,498	871,530	8,351,888
Q3 2003	5,330,160	4,558,251	774,360	10,662,771
Q4 2003	6,077,520	4,148,197	1,006,380	11,232,097
Q1 2004	5,927,220	5,833,758	681,120	12,442,098
Q2 2004	5,886,540	6,537,062	871,530	13,295,132
Q3 2004	6,399,720	5,385,434	524,520	12,309,674
Q4 2004	6,143,580	5,869,232	972,000	12,984,812
Average	5,724,878	4,907,771	871,530	11,504,179
Avg. share (%)	*50*	*43*	*7*	*100*

quarter's minimum value leading to the loyal customer segment size as shown in Table 5.3.

After extracting the loyal customers from the weekly demand by retailer, the remaining demand can be assigned to the smart customer segment, consisting of store-switching and stockpiling customers.

Store-Switching Customer Segment

The store-switching customer segment β_2, like the loyal customer segment, places a fairly constant demand in the market, though variation across quarters is slightly higher. Unlike the loyal customers, the store-switching customers do not prefer a specific retail chain but are well informed about promotions, price sensitive and always purchase at the retailer with the lowest price.

The size of the store-switching segment is determined by analyzing the periods in which only a few promotions are offered. On these occasions, the total demand includes only the loyal and store-switching customers. A split by retailer is obsolete as the segment, by definition, does not have a preferred retail chain. The development of the store-switching customer segment is shown in Table 5.4.

Stockpiling Customer Segment

We define the remaining segment, the stockpiling customer segment β_{1t}, as the weekly aggregated demand in the market less the quarterly constant loyal and store-switching customer segments. The segment shows a high degree of variation across periods, reflecting the characteristics of the stockpiling customer segment β_{1t} as being highly sensitive towards promotion prices and, being well informed, adapts its purchasing behavior according to the retailers promotion strategy. He only enters the market if at least one retailer offers a promotion. A quarterly average of the stockpiling segment is included in the segment overview in Table 5.4.

Given the high volatility the demand of the stockpiling customer segment is the most difficult to forecast. While we restrict ourselves to a numerical decomposition of customer segments, Huchzermeier et al. (2002) estimate the demand of the customer segment with an aggregated stockpiling model. The model provides an excellent fit with $R^2 = 73.8\%$ and thereby allows to forecast the demand of the stockpiling customer with a high accuracy.

Summary

The decomposition of customer demand shows that on average half of the total demand is generated by the loyal customer segment. Another 7% can be attributed to the store-switching customer segment that is well informed about market prices and is constantly present in the market. The remaining large share of 43% forms the stockpiling customer segment. Due to its high demand volatility, the stockpiling customer segment causes variations in weekly demand by a factor of three or more (Fig. 5.10) depending on the promotional activity of the retailers and the resulting competitive pressure. The store-switching customer segment is the smallest segment of the three considered which intuitively suggests that if customers behave in a smart way, most of them will additionally show stockpiling behavior, and restrict their purchases to promotion periods.

Our results for the smart customer segments are in line with the decomposition studies of van Heerde et al. (2004) and Ailawadi et al. (2007) as shown in Table 3.1.

5.3.3 Segmentation-Based Retailer Strategies

Comparing the size of the loyal customer segments of each retail chain with its average weekly demand, we can cluster the six retailers into three different groups: (1) Retailers A and C have a low share of loyal customers with 23% and 36% respectively, which is caused directly by their retail price format being a regular pattern. Customers adapt to the regular promotion pattern and are converted to quasi-loyal stockpiling customers as they return to the same store, in line with the pattern but not on a weekly basis. (2) Retailers B, D and E operate a decentralized HILO promotion

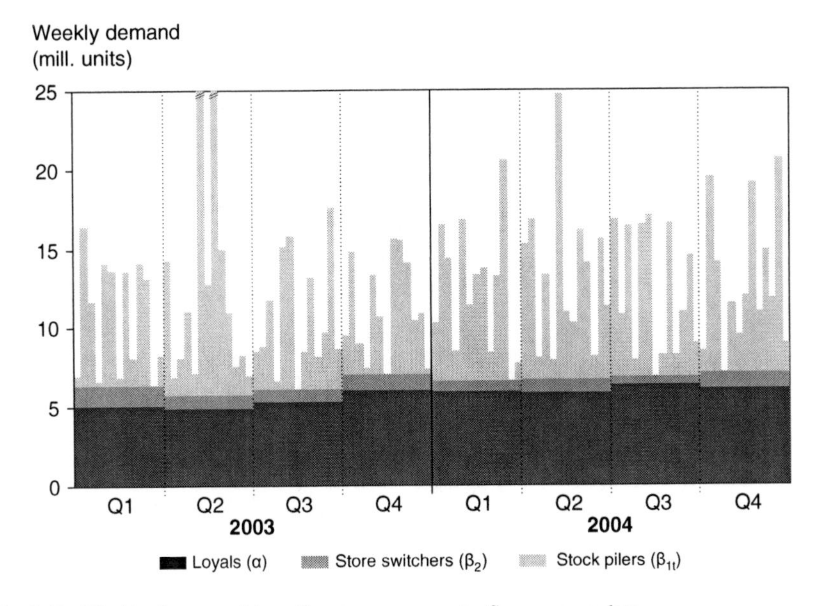

Fig. 5.10 Weekly decomposition of customer segments. Source: own data

strategy and recruit half of their demand from their loyal customers. (3) Retailer F is characterized by a high share of loyal customers amounting to 74%. As his EDLP strategy results in consistently low prices, retailer F attracts a solid high basis of loyal customers. On the other hand, his expectations of gaining large demand peaks from the stockpiling customer segment is limited.

These findings reflect the mixed strategy equilibrium of asymmetric retailers which was derived in Sect. 4.5.1. In order to position the retail chains considered along the y-axis in the market condition matrix in Fig. 4.16, we approximate the ratio of loyal customers of each retail chain $\frac{\alpha_B}{\alpha_A}$ by the reciprocal of their average share of loyal customers $\frac{\alpha_i}{d_{i\,avg}}$ from Table 5.3. The ratio of loyal customers to the entire loyal segment of the competition is not meaningful due to the different dimensions of the retail chains with respect to sizes and hence market shares. A positioning along the x-axis takes place on a weekly basis as the size of the stockpiling segment β_{1t} varies across periods. Note that the bounds shown in Fig. 5.11, τ_i and τ_j, are individual for every retailer.

Retailers A and C with a low share of loyal customers, are positioned at the top end of the market condition matrix. They tend to offer products at a regular price and promote only when the switching customer segment reaches a critical mass. This is achieved by a promotion pattern and a cycle of 3, respectively 4 weeks when the stockpiling customers re-enter the market as quasi-loyal customers to the retailers.

Retailers B, D and E, with a loyal customer share of 50% to 51%, play either regular, promotion or frequently a mixed strategy as a result of their decentralized promotion strategy.

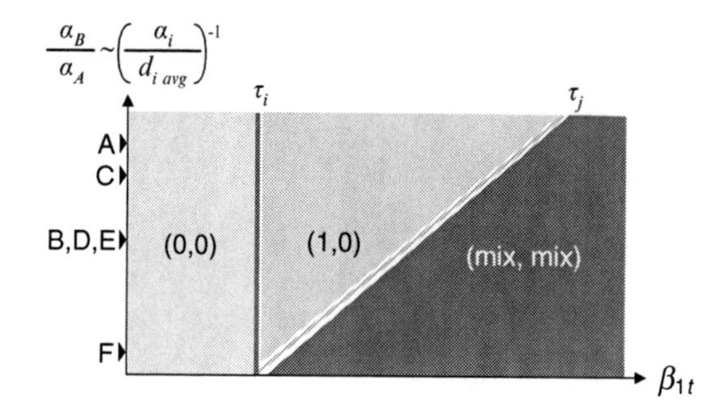

Fig. 5.11 Positioning of retailers in the market condition matrix. Source: own data

The highest share of loyal customers can be observed with retailer F positioned at the bottom of the market condition matrix. As he seems to follow an EDLP strategy, retailer F operates only on one price level with $r = p_h = p_h$ and no bounds τ_F and τ_j exist.

Note that these findings have been drawn from a steady-state market environment where each retailer has adopted his promotion strategy according to his customer base segmentation and vice versa in a given competitive environment.

5.4 Competition Index

The target of a retailer's promotion is to maximize profits by gaining market share and serving customers on a cost efficient level. The demand volatility he faces in the market is caused by customer demand behavior and promotion activities of competitors. The retailer therefore has to base his order quantity along with his promotion decision purely on his expectations of future market conditions. As shown in Sect. 5.3, it is not only his own price that influences customer demand but even more so the price gap between him and the competition. Besides the estimation of future market demand, a prediction of the pricing of his competition enables the retailer to align his own promotion strategy jointly with his order quantity most efficiently.

We therefore introduced the concept of information sharing in Sect. 4.3.3. By sharing downstream information between retailers and manufacturer on promotion schedules in the course of the annual target setting process, manufactures are able to compile the anticipated order quantities of all retailers for future periods. The manufacturer derives the so-called Competition Index based on the aggregated order quantities, reflecting the expected competitive pressure in the market, which in turn is shared downstream with the retailer.

In this section, we will compile the Competition Index based on a second data set containing delivery quantities. We use the Competition Index to predict competitive pressure, and we will discuss its appropriateness and power to support the retailer in his decision-making process.

5.4.1 Order Quantity

We introduce a second data set containing delivery quantities for Pampers Baby Dry for the period July 2002 until December 2004. The data set consists of monthly order quantities aggregated across retailers in Germany.

As the delivery quantity data set is on a monthly basis, its granularity differs from the demand data with respect to the period (month vs. week). In order to merge the information of both data sets, we have to transform either set of data to match the other one's period. As the retailers promote on a weekly basis, we remodel the monthly delivery quantity into a weekly order quantity. The following restrictions therefore have to be considered when interpreting the results of the combined data set: (1) The weekly order quantities are derived from the monthly delivery quantities for each retailer, by weighting each week with its actual demand (ex post). An autocorrelation of order quantity and demand in each period is the result. Hence, a valid analysis of inventory or stockout costs that would result from a mismatch of order quantity, is not possible. (2) Likewise, the prediction of future demand based on the competition would lead to positively forged conclusions.

Taking the above restrictions into account, we focus our interpretation of the Competition Index on the forecasting of competitive pressure.

5.4.2 Competitive Pressure and the Competition Index

Based on the given data set, we compile the Competition Index as an indicator for the competitive pressure of a future period. We therefore define the competitive pressure as the weighted average of competitor promotion frequencies. The competitive pressure thus describes the promoted volume share of all competitors in a given period, and can be interpreted as the promotion frequency across all competing stores.

The Competition Index was defined earlier in Sect. 2.3.3 and is computed as moving average as the ratio of a period's order quantity compared to the average order quantity of past periods. In order to determine the optimal number of previous reference periods, we compute the Competition Index for a range of 1–5 previous periods and compare its correlation to the competitive pressure. The results displayed in Table 5.5 suggest that a reference period of 3 weeks prior to the period in focus is most suited with a correlation of $R^2 = 0.74$.

Table 5.5 Correlation of Competition Index and competitive pressure for various reference periods. Source: own data

Reference periods	R^2
1	0.64
2	0.66
3	0.74
4	0.70
5	0.71

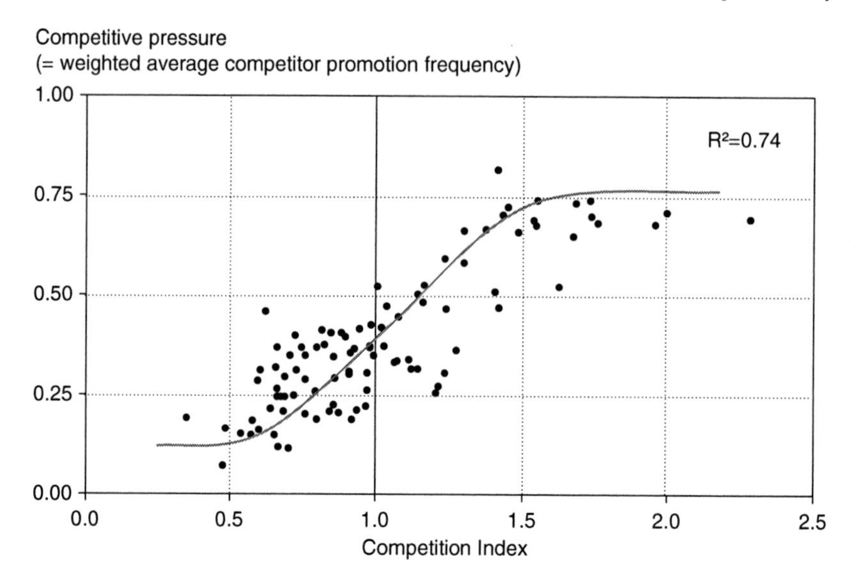

Fig. 5.12 Competition Index and weighted promotion frequency of competition. Source: own data

The Competition Index based on a 3-weeks moving average and the competitive pressure, as the weighted promotion frequency of all competitors, are plotted in Fig. 5.12 for each week of the time period considered from 2003 to 2004 excluding the weeks of the manufacturer promotion (weeks 19–22 in 2003). Each dot in the plot represents 1 week and is defined by the ex ante information from the Competition Index on the x-axis and the ex post result in terms of the competitive pressure of the respective period on the y-axis.

The result is an S-shaped correlation function with a high correlation of R^2 of 0.74. The S-shape reflects the characteristics of the given competitive landscape: due to the decentralized promotion strategies of the three retailers B, D and E, the probability of a competitive pressure being either 0 or 1 or even close to these bounds is negligible.

A high accumulation of incidents can be found in the lower left hand corner where the Competition Index is below 1 and the resulting competitive pressure is below 0.5. With an increasing Competition Index, the competitive pressure increases significantly. Though with fewer events occurring, a Competition Index of >1.5 shows strong evidence of a high competitive pressure with more than 50% of the competitor stores promoting.

In order to use the Competition Index to predict the competitive pressure in a given period, the retailer therefore needs twofold information ex ante: (1) the Competition Index of the respective period as well as (2) the relation of the Competition Index as described by the S-shaped correlation function.

Provided with the Competition Index for a future period by the manufacturer, the retailer is able to predict the competitive pressure he will face in the market. A high competitive pressure indicates a high weighted average promotion frequency of his competitors and in turn a high share of products offered at promotion prices. For example, given the relation from Fig. 5.12, a Competition Index of 1.5 or above would suggest a weighted average promotion frequency in the market of 0.75.

With this information at hand, the retailer is able to improve his forecast accuracy. Unlike in the two-player promotion game, in theory the retailer does not obtain definite information about the competitor's strategy, as he faces an aggregation of several competitors. However, in the course of the two-staged game, he is able to base his promotion decision and order quantity on a more profound prediction of the competitors' promotion strategy. He can therefore increase his supply chain efficiency by reducing inventory as well as stockout cost.

5.5 The Value of Information Sharing

Having characterized the retailers, customers and the information shared, we now determine the value of information sharing numerically. We assume the existence of two symmetric retailers with the characteristics of retailer A and refer to the respective price levels (from Table 5.1) and customer segments (from Table 5.4). Observe that we adapt the size of the store-switching customer segment by retailer A's average market share of 22%. For the stockpiling customer segment, we assume an average size experienced by retailer A in the given period amounting to 5,000,000 customers. The magnitude of effects would increase with the size of the stockpiling customer segment.

Further, we assume for the holding cost h, stockout cost g and cost of goods sold w in line with Proposition 4.4. The individual values are set to mirror the retail environment and are confirmed by retail experts. We set holding cost $h = 0.20$ EUR/unit, cost of goods sold $w = 0.15$ EUR/unit and stockout cost $g = 0.23$ EUR/unit which yields the required condition $g > \frac{(r-w)(r+h)}{(h+w)-(r-w)}$. Finally, we assume $\phi = 0.4$, implying that retailers are to provide a full service level. The variables for the numerical analysis are summarized in Table 5.6. A discussion of manufacturer benefits has already occurred in Sect. 4.4.5.

We begin by considering the magnitude of effects for the retailer and thereafter consider the value of information sharing for the individual customer segments.

5.5.1 *Retailer Benefit*

The value of information sharing for the retailer can be broken down into three different effects: inventory effect, frequency effect and profit effect. These effects

Table 5.6 Definition of variables

Prices	(EUR/unit)	$p_l = 0.17$	$p_h = 0.19$	$r = 0.23$
Cost	(EUR/unit)	$g = 0.23$	$h = 0.20$	$w = 0.15$
Customer	(unit)	$\alpha = 704,138$	$\beta_1 = 5{,}000{,}000$	$\beta_2 = 191{,}736$

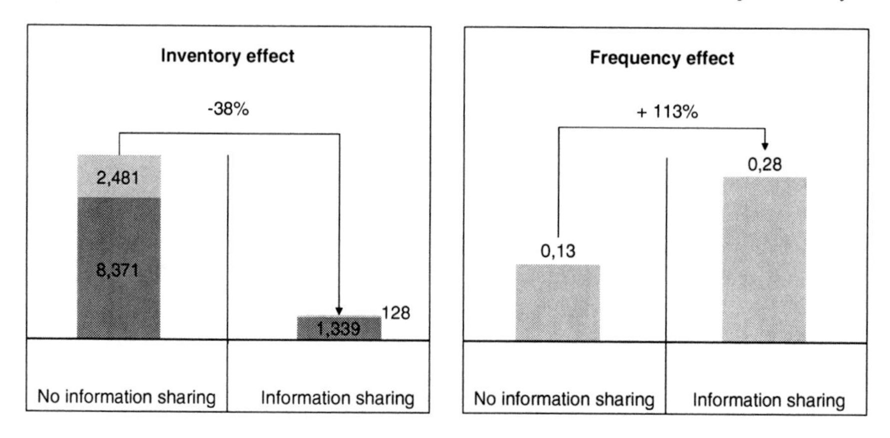

Fig. 5.13 Inventory and frequency effect of retailer A. Source: own data

have been modeled and discussed in Sect. 4.4. In the following, we focus on the numerical results in the following.

Inventory Effect

The inventory effect characterizes the impact of information on the inventory costs. The inventory costs decrease from a total of 10,852 EUR if no information is shared by 38% to 6,691 EUR in information sharing, as shown in Fig. 5.13. The relative effect is larger for the store-switching segment than for the stockpiling segment: while inventory costs due to reduced demand uncertainty from the stockpiling segment are reduced by a high 27% (8,371 EUR to 6,105 EUR), the impact with regard to the store-switching segment is even larger, and reduces inventory costs by 76% (2,481 EUR to 585 EUR). However, due to the larger size of the stockpiling customer segment, the absolute effect is greater for this segment.

Frequency Effect

Given the reduced cost of a promotion, the retailer increases his equilibrium promotion frequency with information sharing as compared with no information sharing. Figure 5.13 shows that the promotion frequency increases from 0.13 to 0.27 if information is available to the retailer. In percentage figures, this implies an increase of over 113% or more than double the initial frequency.

Profit Effect

Despite the large frequency increase, the retailer does not pass along all benefits from information sharing to the customer. Instead, the profit of the retailer increases

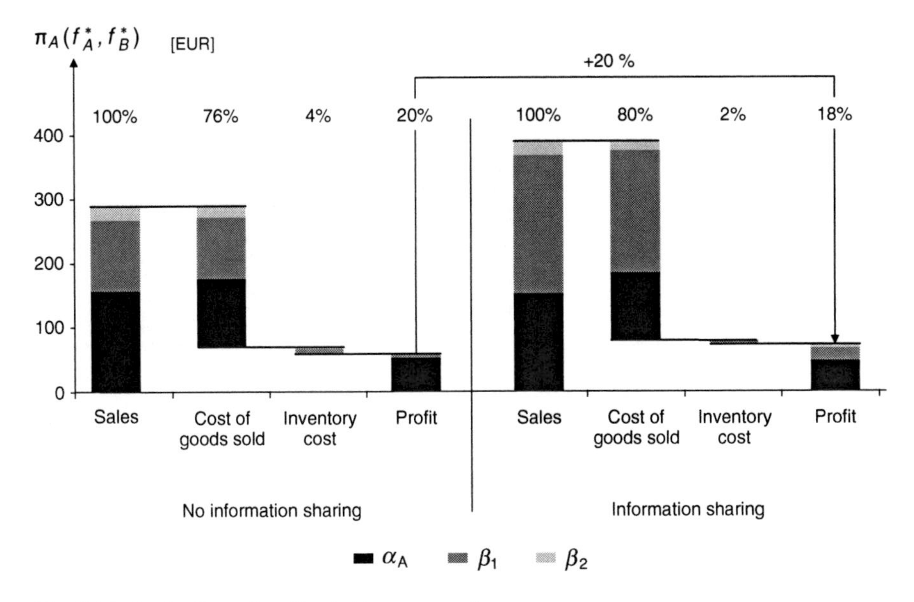

Fig. 5.14 Profit effect of retailer A. Source: own data

with information sharing. In Fig. 5.14, the profit is disaggregated to sales and the respective costs, i.e., wholesale and inventory costs. As evident, sales to the loyal customer segment decrease with information sharing, but the decrease is more than compensated by the larger stockpiling customer segment. With increasing customer base, wholesale costs increase accordingly with information sharing. The major benefit results from the reduced inventory costs. In total, the profit increases by more than 20% from 58,677 EUR with no information sharing to 71,040 EUR with information sharing. Note that these savings are important in the retail industry, where bottom line profits are typically below 2%. A higher absolute profit further contributes to an increase on return on investment (ROI).

5.5.2 *Customer Welfare*

The magnitude of effects is even larger for the customer who derives his benefits from the increased promotion frequency, as shown in Fig. 5.15. The price for the loyal customer segment decreases by 3.1% from over 0.224 EUR/unit with no information sharing to below 0.217 EUR/unit with information sharing. The stockpiling customer segment only attains slight benefits from information sharing, given the already very low price paid with no information sharing. However, the prices he pays with information sharing are lower by 0.2% than with no information sharing. The price reduction is largest for the store-switching customer segments with 5.3% from over 0.218 to 0.207 EUR/unit.

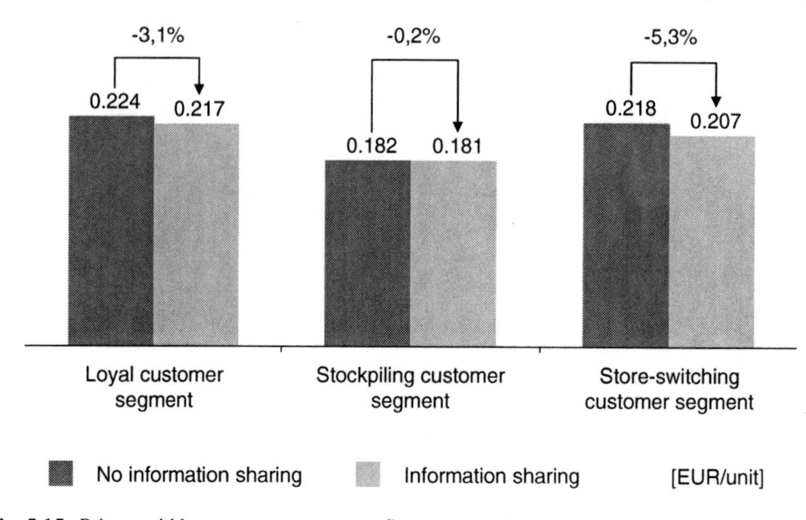

Fig. 5.15 Prices paid by customer segments. Source: own data

With the prices decreasing between 0.2% and 5.3%, all customers pay lower prices if retailers and manufacturers share upstream information on the competitive environment with the Competition Index. The results show that customer welfare increases.

5.5.3 Win–Win–Win Situation

We will summarize the numerical study with a comparison of the value of information sharing for customers and retailers, and extend the results with manufacturer benefits in this environment, as they are evident in literature.

The value of information sharing refers to the performance improvement as a result of an increase in available information. The previous numerical example shows that the retailer's profit increase by more than 20% and customer welfare, depending on the customer segment, increases between 0.2% and 5.3%. Consequently, both retailer and customer have a positive value through information sharing.

The magnitude of benefits from information sharing depends on the size of the stockpiling segment relative to the store-switching and loyal customer segment. Figure 5.16 shows the value of information sharing for the retailer and the three individual customer segments, as a function of the size of the stockpiling customer segment. More explicitly, it shows the percentage decrease of customer prices and the percentage increase of retailer profits.

As evident from the Fig. 5.16, information sharing has no impact on customers and retailers if the size of the stockpiling segment is below its critical size, i.e., $\beta_1 > \tau$. Intuitively, the size of the stockpiling segment is too small to incentivise

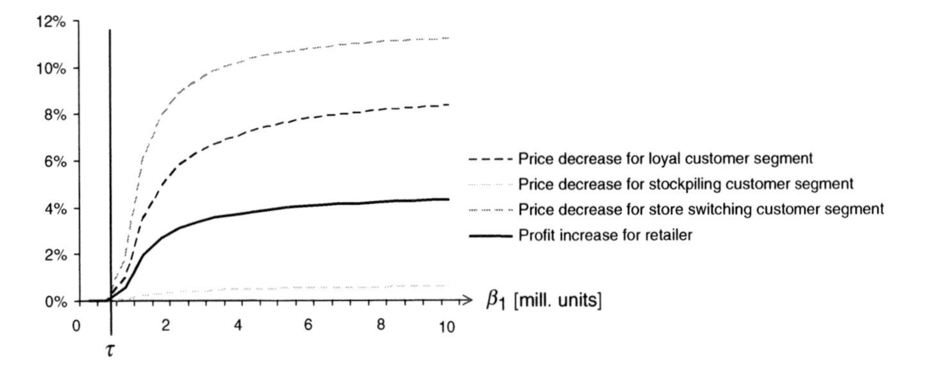

Fig. 5.16 Percentage change for variations of the stockpiling segment. Source: own data

the retailer to offer a promotion, no matter whether he attains information or not. In this situation, information sharing does not change the status quo.

As soon as the size of the stockpiling customer segment exceeds its critical size, information sharing is beneficial for both retailer and customer. Figure 5.16 shows that retailers' profits increase and customer prices decrease with information sharing. Moreover, we see a saturation of the value of information sharing for an increasing size of the stockpiling segment. To summarize: the magnitude of benefits for the retailer and for the customers depends on the market condition.

Finally, let us take these effects and compare them to the benefits the manufacturer can derive from information sharing. As shown by Iyer and Ye (2000), the manufacturer derives benefits from the downstream information on the retailer's promotion schedule. Moreover, the manufacturer derives benefits from lower customer prices in competition with other brands. In total, we can therefore validate the existence of a win–win–win situation due to sharing of upstream information in terms of the Competition Index.

Chapter 6
Conclusion and Managerial Insights

This thesis considers the collaboration between manufacturer and retailer through sharing upstream information that is suited to serve as the basis of a retailer's joint optimization of promotion strategy and order quantity. In a competitive environment, we apply the fundamentals of game theory to describe the retailer's decision-making process in a two-stage approach and determine the value of information sharing. The Competition Index is introduced and provides a new and unique framework for upstream information sharing. We further extend our view to the entire supply chain including manufacturer, retailer as well as loyal and smart customer segments. As a result, we demonstrate the possibility of improving supply chain efficiency and thereby achieving a win–win–win situation by means of collaborative promotions.

We will summarize the key findings by pointing out (1) the impact of customer demand and (2) the resulting strategic options for retailers. We suggest as an appropriate strategy the idea of (3) improving supply chain efficiency by collaborative promotions.

6.1 Impact of Customer Demand

Customer demand in promotions is highly volatile, causing the majority of stockouts or excess inventory. We must first understand the causes for volatility in a segmented customer base and then determine how collaboration can increase supply chain efficiency by reducing inventory costs.

To underline the theory, we have provided an empirical decomposition based on retail level data. We have addressed the heterogeneous customer base and derived an appropriate segmentation for our problem. We have segmented the customers into loyal and smart customers, with a further split of smart customers into stockpiling and store-switching customers. Although retail data do not allow us to model the disaggregate consumer response due to price promotions, we have shown how to obtain substantively relevant decomposition effects.

On average, across the six retailers observed, we found an equal split of the market between loyal and smart customers: 50% of the diapers customers are loyal to

D. Wiehenbrauk, *Collaborative Promotions*, Lecture Notes in Economics and Mathematical Systems 643, DOI 10.1007/978-3-642-13393-0_6,

their preferred retailer, whereas the other 50% behave in a smart way and switch between retailers to get a good deal. Moreover, the 50% smart customers are primarily, with 43%, also those who further optimize their household budget by stockpiling behavior.

With an understanding of the size of these customer segments, we provide insight into the sources of customer demand volatility in the diapers category. Moreover, we show how customers can be segmented in the diapers category in line with a given retailer strategy.

Customer segmentation is unique to every retailer in the market. Driven by his retail concept, number of stores, locations and last but not least, pricing, each retailer faces a different mix of loyal, store-switching and stockpiling customers. Our empirical applications show how the decomposition results are moderated by characteristics of the price promotion. We find that the distribution of customer segments at each individual retailer clearly depends on the retailer's strategy and, vice versa, retailers require different strategies depending on the segmentation of the customer base they face.

6.2 Strategic Options for Retailers

We have modeled a marketing-manufacturing interface problem in which the retailer coordinates his promotion and order decisions. Thereby, we have taken into account that promotion decisions have a significant effect on retailer's order quantities. The optimization of the retailer's decision was modeled with respect to order quantity, by means of a newsvendor problem and with respect to promotion frequency, by a Bertrand Nash equilibrium. Thereby, the mixed strategy equilibrium was derived based on the expected profits, and the resulting best response correspondence for symmetric retailers.

We introduced different price levels for regular and promotion pricing as well as the frequency and depth of a promotion. In the empirical data analysis we observed combinations of these basic promotion elements, and derived the individual retail price formats.

In order to provide a meaningful interpretation of the different strategies, we extended the model to asymmetric retailers and derived the market condition matrix.

Retailers have to decide on prices and inventory across their stores. These two decisions need to match the underlying promotion strategy that is reflected in the retail price format. Three fundamentally different pricing strategies have been derived from the basic promotion elements and identified in practice: EDLP, HILO and pattern.

For the retailer with a high share of loyal customers relative to the size of the smart customer segment, an EDLP strategy is the optimal strategy. By not promoting, he exploits profits from his loyal customer segment. Retailers with a lower share of loyal customers rely on promotions in order to attract the smart customer segments and to increase profits. As soon as the size of the stockpiling segment is large enough, these retailers offer a promotion. With a pattern strategy, retailers actively influence the market by artificially increasing the size of the stockpiling segment by making themselves predictable to the customer. Whereas with a HILO strategy, retailers rely on market conditions.

Clearly, the overall benefit for the retailer depends on the match of his promotion and pricing strategy with his customer base. When considering a mature market, we find that different strategies and retail price formats are necessary, based on the individual customer segmentation of each retailer. Retailers with a lower share of loyal customers may find it beneficial to employ a pattern; those with a medium share of loyals may use HILO, and those with a high share of loyals, EDLP. We demonstrate the market positioning of retailers by means of their underlying pricing strategy in a unique market condition matrix.

6.3 Improving Supply Chain Efficiency by Collaborative Promotions

When deriving upstream information as a means to improve supply chain efficiency, the collaboration of manufacturer and retailer is key. Each party holds valuable private information that, if shared, can be of common benefit. While retailers have a defined promotion strategy, and therefore an anticipated promotion schedule, manufacturers have the ability to aggregate the individual retailer schedules to a common set, if shared.

Based on the aggregated knowledge about future promotion plans and order quantities of individual retailers, manufacturers are able to compile the order quantity of future periods and compute the Competition Index. Due to its high correlation the Competition Index represents a powerful ex ante indicator on competitive pressure in a given period. Sharing this information with the retailer enables him to optimize his promotion strategy and order quantity.

In order to determine the value of information sharing, we have set up two different scenarios. The first scenario, no information sharing, describes the status quo, where retailers base their order quantity and promotion decision on their expectations of competitor's behavior. In the second scenario, we considered the impact of sharing the upstream information on the retailer's decisions, and further analyzed the impact on both retailer and customer.

When comparing the two scenarios, we find that sharing the Competition Index has the potential to dramatically improve supply chain efficiency. From the information of the Competition Index, a retailer can conclude on whether his competitor offers a product at the regular or promotion price. With reduced uncertainty on competitor's pricing, the demand uncertainty is reduced in turn. Lower demand uncertainty allows to better fit supply and demand and thereby reduced inventory costs, which we confirmed by providing evidence of a positive inventory effect. In a numerical example we showed that inventory costs of a promotion can decrease by a high 38% with information sharing.

If the cost of a promotion decreases, it is profit maximizing for a retailer to increase his promotion frequency, resulting in the frequency effect. The frequency effect can be dramatic with the frequency tripling on information sharing as compared to without information sharing as evident from the numerical study.

The impact of the reduced inventory costs and the resulting frequency effect is twofold. Firstly, the retailer attains higher profits. The profit increased by more than 20% in our numerical study, which reveals its significance when set in relation to the typically low margins in retail industry. Secondly, customers derive their benefits from an increased promotion frequency, which results in lower prices compared to a non-collaborative environment. We were able to show that customer prices decreased between 0.2% and 5.3%, depending on the observed customer segment.

Finally, the manufacturer benefits from collaborative promotions primarily due to the lower customer prices, allowing him to regain market share from competition and to close the price gap to private labels. Further, with early information on retailer promotions, the manufacturer can intuitively smoothen production planning.

The magnitude of effects, i.e., the profit increase for the retailer and the welfare increase for the customer, depends on the size of the three customer segments. The more stockpiling customers are in the market, the higher the benefits from information sharing for the retailer, and the higher the welfare effect for the customer, and finally, the higher the benefits for the manufacturer.

With the Competition Index we have identified a powerful piece of information that is not opposing to competition law. As shown in theory as well as in the numerical example, the retailer only benefits from information sharing if the promotion frequency increases since inventory and frequency effect are linked and occur at the same time. Therefore a benefit for the retailer is linked to an improvement of customer welfare. The customer does not pay for the retailer's increasing profit, which is funded by an improvement of supply chain efficiency. So the final effect of information sharing is positive for all parties involved, leading to a win–win–win situation.

Collaborative Promotions by sharing upstream information compiled in the Competition Index increase supply chain efficiency by eliminating inventory. The retailer is enabled to disentangle the doom loop of promotions by collaborating with the manufacturer. The Competition Index provides a new and unique framework for upstream information sharing and generates a broad win–win–win situation for customers, retailers and manufactures.

Appendix A
Proofs and Calculations

In this Appendix, we provide the calculations for the proofs of the best response on no information sharing and information sharing. The following transformations are all based on the following equations, which are derived from the critical fractile in (4.3).

$$
\begin{aligned}
c(p_l) &= \frac{c_u(p_h)}{c_u(p_h)+c_o}, & c_u(p_l) &= p_l + g - w,\\
c(p_h) &= \frac{c_u(p_h)}{c_u(p_h)+c_o}, & c_u(p_h) &= p_h + g - w,\\
& & c_u(r) &= r + g - w,\\
c(r) &= \frac{c_u(r)}{c_u(r)+c_o}, & c_o &= h + w.
\end{aligned} \tag{A.1}
$$

We use the three step procedure as described in Sect. 4.2.2.2 for each of the $k = 8$ cases in the scenario of no information sharing and the $k = 3$ cases in the scenario of information sharing. Further, we remain with the assumption that $\alpha > \beta_2$ as described in Sect. 4.1.3 and for the stockout cost g, the condition $g(1-c(r)) > r-w$ must hold true.

A.1 Best Response for the Scenario No Information Sharing

Proposition A.1. $\forall g(1-c(r)) > r-w$, *the best response of retailer A for the scenario of no information sharing* $\hat{\delta}_{Ak}(\hat{f}_B, \beta_{1t})$ *depends on the size of the stockpiling segment* β_{1t} *and the competitor's promotion frequency* $\hat{f}_B$. *The critical size for the stockpiling segment* τ *is derived as*

$$\tau = \frac{\alpha \Delta \Upsilon - \beta_2 \Delta \hat{\Lambda}_8}{\hat{\Lambda}_8(p)}.$$

We attain two different cases

Case i: *If $\beta_{1t} < \tau$ or $\hat{f}_B \notin [0, c(p_h)]$, then $\hat{\delta}_{Ak}(\hat{f}_B, \beta_{1t}) < 0 \; \forall \hat{f}_B \in [0,1]$ and retailer A's best response is to play his pure strategy "regular" independent of retailer B's strategy. He has a dominant best response:*

$$\hat{f}_A^* = 0.$$

Case ii: *If $\beta_{1t} > \tau$ and $\hat{f}_B \in [0, c(p_h)]$, retailer A's best response is to play strategically and mix between his pure strategies "regular" and "promotion" with probability*

$$\hat{f}_A^* = \begin{cases} 1 & \text{if } \hat{f}_B \in [0, \hat{f}_B^\dagger), \\ [0,1] & \text{if } \hat{f}_B = \hat{f}_B^\dagger, \\ 0 & \text{if } \hat{f}_B \in (\hat{f}_B^\dagger, 1], \end{cases}$$

with

$$\hat{f}_B^\dagger = \frac{-\alpha\Delta\Upsilon + \beta_{1t}\hat{\Lambda}_8(p) + \beta_2\Delta\hat{\Lambda}_8}{\beta_{1t}\hat{\Omega}_8(p) + \beta_2\Delta\hat{\Omega}_8}.$$

For the proof of the proposition, we shall apply described three step procedure to each of the $k = 8$ cases.

Observe that we require for the stockout cost $g(1-c(r)) > r-w$. If this condition is fulfilled, then also $g(1-c(p_h)) > p_h - w$ and $g(1-c(p_l)) > p_l - w$ hold true.

Case 1 $\hat{f}_B \in \left[\frac{c(p_h)}{\phi}, 1\right], \phi \in [c(p_h), 1]$

1. $\hat{\delta}_{A1}(\hat{f}_B = 0) = -\alpha\Delta\Upsilon + \beta_{1t}\hat{\Lambda}_1(p) + \beta_2\Delta\hat{\Lambda}_1$ with

$$\begin{aligned}
\hat{\Lambda}_1(p) &= \phi p_l + (1-\phi)p_h - w - \phi\frac{c_u(p_l)}{2} - (1-\phi)c_u(p_h) \\
&= \phi\frac{c_u(p_l)}{2} - g, \\
\Delta\hat{\Lambda}_1 &= \phi p_l + (1-\phi)p_h - \frac{r}{2} - \frac{w}{2} - \phi\frac{c_u(p_l)}{2} - (1-\phi)c_u(p_h) + \frac{c_u(r)}{2} \\
&= \frac{1}{2}\left(\phi c_u(p_l) - g\right),
\end{aligned}$$

where

$$\begin{aligned}
&\hat{\Lambda}_1(p) \quad < 0 \quad \forall\phi \in [c(p_h), 1], \\
&\Delta\hat{\Lambda}_1 \begin{cases} < 0 & \forall\phi \in \left[c(p_h), \frac{g}{c_u(p_l)}\right], \\ > 0 & \forall\phi \in \left[\frac{g}{c_u(p_l)}, 1\right], \end{cases}
\end{aligned}$$

$\forall\phi < \frac{g}{c_u(p_l)}$, both $\hat{\Lambda}_1(p)$ and $\Delta\hat{\Lambda}_1$ are negative and the best response has a negative sign, i.e., $\delta_{A1}(\hat{f}_B = 0) < 0$.
$\forall\phi > \frac{g}{c_u(p_l)}$, the term $\Delta\hat{\Lambda}_1$ is positive and in order to confirm the best response has a negative sign, we need to impose the following constraint on the size of the stockpiling segment β_{1t}

$$\beta_{1t} > \frac{\alpha\Delta\Upsilon - \frac{\beta_2}{2}(\phi c_u(p_l) - g)}{\phi\frac{c_u(p_l)}{2} - g}$$

with a positive numerator (because $\alpha\Delta\Upsilon > \frac{\beta_2}{2}(\phi c_u(p_l) - g)$) and a negative denominator. Consequently, the condition is always fulfilled and we can conclude that

$$\hat{\delta}_{A1}(\hat{f}_B = 0) < 0, \quad \forall\phi \in [c(p_h), 1] \tag{A.2}$$

the CHANCE of retailer A's best response is negative.

2. $\hat{\delta}_{A1}(\hat{f}_B = 1) = -\alpha\Delta\Upsilon + \beta_{1t}\big(\hat{\Lambda}_1(p) - \hat{\Omega}_1(p)\big) + \beta_2\big(\Delta\hat{\Lambda}_1 - \Delta\hat{\Omega}_1\big)$, where

$$\hat{\Lambda}_1(p) - \hat{\Omega}_1(p) = \frac{1}{2}(\phi c_u(p_l) - g),$$
$$\Delta\hat{\Lambda}_1 - \Delta\hat{\Omega}_1 = \frac{1}{2}(\phi c_u(p_l) - g),$$

where

$$\hat{\Lambda}_1(p) - \hat{\Omega}_1(p) = \Delta\hat{\Lambda}_1 - \Delta\hat{\Omega}_1 \begin{cases} < 0 \ \forall\phi \in \left[c(p_h), \frac{g}{c_u(p_l)}\right], \\ > 0 \ \forall\phi \in \left[\frac{g}{c_u(p_l)}, 1\right], \end{cases}$$

$\forall\phi < \frac{g}{c_u(p_l)}$ both $\hat{\Lambda}_1(p) - \hat{\Omega}_1(p)$ and $\Delta\hat{\Lambda}_1 - \Delta\hat{\Omega}_1$ are negative and consequently the best response is negative.
$\forall\phi > \frac{g}{c_u(p_l)}$ both $\hat{\Lambda}_1(p) - \hat{\Omega}_1(p)$ and $\Delta\hat{\Lambda}_1 - \Delta\hat{\Omega}_1$ are positive and attain a bound for the size of the stockpiling segment at

$$\beta_{1t} < \frac{\Delta\Upsilon - \frac{\beta_2}{2}\big(\phi c_u(p_l) - g\big)}{\phi c_u(p_l) - g},$$

which is always fulfilled due to a small denominator and consequently a large fraction which will not be exceeded by the size of the switching segment. Consequently, we can conclude that

$$\hat{\delta}_{A1}(\hat{f}_B = 1) < 0, \quad \forall\phi \in [c(p_h), 1]. \tag{A.3}$$

The BASE and CHANCE elements are not large enough to balance the negative risk.

3. Best response $\hat{f}_1^*$
   From (A.2) and (A.3), one can conclude that the best response has a negative sign for the complete domain of $\hat{f}_B \in \left[\frac{c(p_h)}{\phi}, 1\right]$. Thus, the retailer's best response in Case 1 is
$$\hat{f}_1^* = 0.$$

Case 2 $\hat{f}_B \in \left[\frac{c(p_h)}{\phi}, c(r)\right], \phi \in \left[\frac{c(p_h)}{c(r)}, 1\right]$

1. $\delta_{A2}(\hat{f}_B = 0) = -\alpha\Delta\Upsilon + \beta_{1t}\hat{\Lambda}_2(p) + \beta_2\Delta\hat{\Lambda}_2$ with

$$\begin{aligned}\hat{\Lambda}_2(p) &= \phi p_l + (1-\phi)p_h - w - \phi\frac{c_u(p_l)}{2} - (1-\phi)c_u(p_h)\\ &= \phi\frac{c_u(p_l)}{2} - g,\\ \Delta\hat{\Lambda}_2 &= \phi p_l + (1-\phi)p_h - \frac{r}{2} - \frac{w}{2} - \phi\frac{c_u(p_l)}{2} - (1-\phi)c_u(p_h)\\ &= \frac{1}{2}\big(\phi c_u(p_l) - c_u(r) - g\big),\end{aligned}$$

   where

$$\begin{aligned}\hat{\Lambda}_2(p) &< 0, \quad \forall\phi \in \left[\frac{c(p_h)}{c(r)}, 1\right],\\ \Delta\hat{\Lambda}_2 &< 0, \quad \forall\phi \in \left[\frac{c(p_h)}{c(r)}, 1\right].\end{aligned}$$

   Consequently, the best response has a negative sign at $\hat{f}_B = 0$

$$\delta_{A2}(\hat{f}_B = 0) < 0, \quad \forall\phi \in \left[\frac{c(p_h)}{c(r)}, 1\right]. \tag{A.4}$$

2. $\delta_{A2}(\hat{f}_B = c(r)) = -\alpha\Delta\Upsilon + \beta_{1t}\big(\hat{\Lambda}_2(p) - c(r)\hat{\Omega}_2(p)\big) + \beta_2\big(\Delta\hat{\Lambda}_2 - c(r)\Delta\hat{\Omega}_2\big)$ with

$$\begin{aligned}\hat{\Lambda}_2(p) - c(r)\hat{\Omega}_2(p) &= \frac{1}{2}\big(\phi c_u(p_l) - g(2 - c(r))\big),\\ \Delta\hat{\Lambda}_2 - c(r)\Delta\hat{\Omega}_2 &= \frac{1}{2}\big(\phi c_u(p_l) - g\big),\end{aligned}$$

$$\begin{aligned}\hat{\Lambda}_2(p) - c(r)\hat{\Omega}_2(p) &< 0, \quad \forall\phi \in \left[\frac{c(p_h)}{c(r)}, 1\right],\\ \Delta\hat{\Lambda}_2 - c(r)\Delta\hat{\Omega}_2 &< 0, \quad \forall\phi \in \left[\frac{c(p_h)}{c(r)}, 1\right],\end{aligned}$$

and we can conclude

$$\hat{\delta}_{A2}(\hat{f}_B = 0) < 0, \quad \forall \phi \in \left[\frac{c(p_h)}{c(r)}, 1\right]. \tag{A.5}$$

3. Best response $\hat{f}_2^*$ As evident from (A.4) and (A.5) the best response has a negative sing for the complete domain of $\hat{f}_B \in \left[\frac{c(p_h)}{\phi}, c(r)\right]$. Thus, the retailer' best response in Case 2 is

$$\hat{f}_2^* = 0.$$

Case 3 $\hat{f}_B \in \left[\frac{c(p_l)}{\phi}, c(p_h)\right], \phi \in \left[\frac{c(p_l)}{c(p_h)}, 1\right]$.

Observe that in this case $\phi > \frac{c(p_l)}{c(p_h)}$ and consequently $\phi \to 1$

1. $\hat{\delta}_{A3}(\hat{f}_B = 0) = -\alpha \Delta \Upsilon + \beta_{1t} \hat{\Lambda}_3(p) + \beta_2 \Delta \hat{\Lambda}_3$

$$\begin{aligned} \hat{\Lambda}_3(p) &= \phi p_l + (1-\phi) p_h - w - \phi \frac{c_u(p_l)}{2}, \\ \Delta \hat{\Lambda}_3(p) &= \phi p_l + (1-\phi) p_h - \frac{w}{2} - \frac{r}{2} - \phi \frac{c_u(p_l)}{2}, \end{aligned}$$

$$\begin{aligned} \hat{\Lambda}_3(p) &< 0, \quad \forall \phi \in \left[\frac{c(p_l)}{c(p_h)}, 1\right], \\ \Delta \hat{\Lambda}_3(p) &< 0, \quad \forall \phi \in \left[\frac{c(p_l)}{c(p_h)}, 1\right]. \end{aligned}$$

Consequently, the best response has a negative sign at $\hat{f}_B = 0$

$$\hat{\delta}_{A3}(\hat{f}_B = 0) < 0, \quad \forall \phi \in \left[\frac{c(p_l)}{c(p_h)}, 1\right]. \tag{A.6}$$

2. $\hat{\delta}_{A3}(\hat{f}_B = c(p_h)) = -\alpha \Delta \Upsilon + \beta_{1t}\big(\hat{\Lambda}_3(p) - c(p_h)\hat{\Omega}_3(p)\big) + \beta_2\big(\Delta \hat{\Lambda}_3 - c(p_h)\Delta \hat{\Omega}_3\big)$, where

$$\begin{aligned} \hat{\Lambda}_3(p) - c(p_h)\hat{\Omega}_3(p) &= \phi p_l + (1-\phi) p_h - w - \phi \frac{c_u(p_l)}{2} - \frac{c(p_h)}{2} \\ &\quad \times \big(\phi^2 p_l + (1-\phi^2) p_h - w - \phi^2 c_u(p_l) + (1-\phi^2) c_o\big), \\ \Delta \hat{\Lambda}_3 - c(p_h)\Delta \hat{\Omega}_3 &= \phi p_l + (1-\phi) p_h - \frac{r}{2} - \frac{w}{2} - \phi \frac{c_u(p_l)}{2} - \frac{c(p_h)}{2} \\ &\quad \times \big(\phi^2 p_l + (1-\phi^2) p_h - r - \phi^2 (c_u(p_l) + c_o)\big). \end{aligned}$$

The best response function in Case 3 is convex in ϕ as given by

$$\frac{\partial^2 \hat{\delta}_{A3}(\hat{f}_B = c(p_h))}{\partial \phi^2} = c(p_h)(\beta_{1t} + \beta_2)(p_h - p_l + c_u(p_l) + c_o) > 0.$$

Consequently, showing that the best response is negative can be achieved by showing that the best response is negative at either bound $\phi = 0$ and $\phi = 1$. Applying the transformations from (A.1), the best response at the bounds is given by

$$\hat{\delta}_{A3}\big(\hat{f} = c(p_h), \phi = 0\big) = -\alpha(r - p_h) - \beta_{1t}\Big(-(p_h - w) + \frac{c(p_h)}{2}(p_h - w + c_o)\Big)$$
$$-\frac{\beta_2}{2}(r - p_h)(1 - c(p_h)) + \frac{\beta_2}{2}(p_h - w),$$

with $\alpha(r - p_h) > \frac{\beta_2}{2}(p_h - w)$ and $\beta_{1t}\big(-(p_h - w) + \frac{c(p_h)}{2}(p_h - w + c_o)\big) > 0$, $\forall p_h - w < g(1 - c(p_h))$ we find that $\hat{\delta}_{A3}\big(\hat{f} = c(p_h), \phi = 0\big) < 0$.

$$\hat{\delta}_{A3}\big(\hat{f} = c(p_h), \phi = 1\big) = -\alpha(r - p_l) - \frac{\beta_{1t}}{2}\big(-(p_l - w) + g(1 - c(p_h))\big)$$
$$-\frac{\beta_2}{2}\big(c_u(r) - c(p_h)(c_u(r) + c_o)\big) + \frac{\beta_2}{2}(p_l - w),$$

with $\alpha(r - p_l) > \frac{\beta_2}{2}(p_l - w)$ and $\frac{\beta_{1t}}{2}\big(-(p_l - w) + g(1 - c(p_h))\big) > 0$, $\forall p_l - w < g(1 - c(p_h))$ we find that $\hat{\delta}_{A3}\big(\hat{f} = c(p_h), \phi = 1\big) < 0$.
We can conclude

$$\hat{\delta}_{A3}(\hat{f} = c(p_h)) < 0, \quad \forall \phi \in \left[\frac{c(p_l)}{c(p_h)}, 1\right]. \tag{A.7}$$

3. Best response $\hat{f}_3^*$ From (A.6) and (A.7) we find the best response of retailer A in Case 3 as

$$\hat{f}_3^* = 0, \quad \forall \hat{f}_B \in \left[\frac{c(p_l)}{\phi}, c(p_h)\right].$$

Case 4 $\hat{f}_B \in [c(p_h), c(r)], \phi \in \left[\frac{c(p_l)}{c(r)}, 1\right]$

1. $\hat{\delta}_{A4}(\hat{f}_B = 0) = -\alpha \Delta \Upsilon + \beta_{1t} \hat{\Lambda}_4(p) + \beta_2 \Delta \hat{\Lambda}_4$

$$\hat{\Lambda}_4(p) = \phi p_l + (1 - \phi) p_h - w - \phi \frac{c_u(p_l)}{2} - (1 - \phi)\frac{c_u(p_h)}{2}$$
$$= \frac{1}{2}\big(\phi p_l + (1 - \phi) p_h - w - g\big),$$

$$\Delta\hat{\Lambda}_4 = \phi p_l + (1-\phi)p_h - \frac{r}{2} - \frac{w}{2} - \phi\frac{c_u(p_l)}{2} - (1-\phi)\frac{c_u(p_h)}{2}$$
$$= \frac{1}{2}(p_h - w - c_u(r)),$$

$$\hat{\Lambda}_4(p) < 0, \quad \forall\phi \in \left[\frac{c(p_l)}{c(r)}, 1\right],$$
$$\Delta\hat{\Lambda}_4 \ < 0, \quad \forall\phi \in \left[\frac{c(p_l)}{c(r)}, 1\right].$$

Consequently, the best response has a negative sign at $\hat{f}_B = 0$

$$\hat{\delta}_{A4}(\hat{f}_B = 0) < 0, \quad \forall\phi \in \left[\frac{c(p_l)}{c(r)}, 1\right]. \tag{A.8}$$

2. $\hat{\delta}_{A4}(\hat{f}_B = c(r)) = -\alpha\Delta\Upsilon + \beta_{1t}\big(\hat{\Lambda}_4(p) - c(r)\hat{\Omega}_4(p)\big) + \beta_2\big(\Delta\hat{\Lambda}_4 - c(r)\Delta\hat{\Omega}_4\big)$, where

$$\hat{\Lambda}_4(p) - c(r)\hat{\Omega}_4(p) = \phi p_l + (1-\phi)p_h - w - \phi\frac{c_u(p_l)}{2} - (1-\phi)\frac{c_u(p_h)}{2}$$
$$-\frac{c(r)}{2}\big(\phi^2 p_l + (1-\phi^2)p_h - w - \phi^2 c_u(p_l)$$
$$-(1-\phi)(c_u(p_h) - \phi c_o)\big),$$
$$\Delta\hat{\Lambda}_4 - c(r)\Delta\hat{\Omega}_4 = \phi p_l + (1-\phi)p_h - \frac{r}{2} - \frac{w}{2} - \phi\frac{c_u(p_l)}{2} - (1-\phi)\frac{c_u(p_h)}{2}$$
$$-\frac{c(r)}{2}\big(\phi^2 p_l + (1-\phi^2)p_h - r - \phi^2 c_u(p_l)$$
$$-(1-\phi)c_u(p_h) - c_o(1-\phi(1-\phi))\big).$$

The best response function in Case 4 is convex in ϕ as given by

$$\frac{\partial^2\hat{\delta}_{A4}(\hat{f}_B = c(r))}{\partial\phi^2} = c(r)(\beta_{1t} + \beta_2)(p_h - p_l + c_u(p_l) + c_o) > 0.$$

Consequently, showing that the best response is negative can be achieved by showing that the best response is negative at either bound $\phi = 0$ and $\phi = 1$. Applying the transformations from (A.1), the best response at the bounds is given by

$$\hat{\delta}_{A4}\big(\hat{f} = c(r), \phi = 0\big) = -\alpha(r - p_h) - \frac{\beta_{1t}}{2}\big(-(p_h - w) + g(1 - c(r))\big)$$
$$+\frac{\beta_2}{2}(p_h - w),$$

with $\alpha(r - p_h) > \frac{\beta_2}{2}(p_h - w)$ and $\frac{\beta_{1t}}{2}\big(-(p_h - w) + g(1 - c(r))\big) > 0, \forall p_h - w < g(1 - c(r))$ we find that $\hat{\delta}_{A4}\big(\hat{f} = c(r), \phi = 0\big) < 0$.

$$\hat{\delta}_{A4}\big(\hat{f} = c(r), \phi = 1\big) = -\alpha(r - p_l) - \frac{\beta_{1t}}{2}\big(-(p_l - w) + g(1 - c(r))\big) + \frac{\beta_2}{2}(p_l - w),$$

with $\alpha(r - p_l) > \frac{\beta_2}{2}(p_l - w)$ and $\frac{\beta_{1t}}{2}\big(-(p_l - w) + g(1 - c(r))\big) > 0, \forall p_l - w < g(1 - c(p_h))$ we find that $\hat{\delta}_{A4}\big(\hat{f} = c(r), \phi = 1\big) < 0$.
We can conclude

$$\hat{\delta}_{A4}\big(\hat{f} = c(r)\big) < 0, \quad \forall \phi \in \left[\frac{c(p_l)}{c(r)}, 1\right]. \tag{A.9}$$

3. Best response $\hat{f}_4^*$ Given the results of (A.8) and (A.9) the best response of retailer A in Case 4 is
$$\hat{f}_4^* = 0, \quad \forall \hat{f}_B \in [c(p_h), c(r)].$$

Case 5 $f \in [c(r), 1]$ $\phi \in \left[c(p_l), \frac{c(p_h)}{c(r)}\right]$

1. $\hat{\delta}_{A5}(\hat{f}_B = 0) = -\alpha \Delta \Upsilon + \beta_{1t} \hat{\Lambda}_5(p) + \beta_2 \Delta \hat{\Lambda}_5$

$$\begin{aligned}
\hat{\Lambda}_5(p) &= \phi p_l + (1 - \phi) p_h - w - \phi \frac{c_u(p_l)}{2} - (1 - \phi) \frac{c_u(p_h)}{2} \\
&= \frac{1}{2}\big(\phi p_l + (1 - \phi) p_h - w - g\big), \\
\Delta \hat{\Lambda}_5 &= \phi p_l + (1 - \phi) p_h - \frac{r}{2} - \frac{w}{2} - \phi \frac{c_u(p_l)}{2} - (1 - \phi) \frac{c_u(p_h)}{2} + \frac{c_u(r)}{2} \\
&= \frac{1}{2}(\phi p_l + (1 - \phi) p_h - w),
\end{aligned}$$

$$\begin{aligned}
\hat{\Lambda}_5(p) &< 0, \quad \forall \phi \in \left[c(p_l), \frac{c(p_h)}{c(r)}\right], \\
\Delta \hat{\Lambda}_5 &> 0, \quad \forall \phi \in \left[c(p_l), \frac{c(p_h)}{c(r)}\right].
\end{aligned}$$

With $\alpha > \beta_2$, we can conclude that the best response has a negative sign at $\hat{f}_B = 0$

$$\hat{\delta}_{A5}(\hat{f}_B = 0) < 0, \quad \forall \phi \in \left[c(p_l), \frac{c(p_h)}{c(r)}\right]. \tag{A.10}$$

2. $\hat{\delta}_{A5}(\hat{f}_B = c(r)) = -\alpha\Delta\Upsilon + \beta_{1t}\big(\hat{\Lambda}_5(p) - \hat{\Omega}_5(p)\big) + \beta_2\big(\Delta\hat{\Lambda}_5 - \Delta\hat{\Omega}_5\big)$, where

$$\begin{aligned}
\hat{\Lambda}_5(p) - \hat{\Omega}_5(p) &= \phi p_l + (1-\phi)p_h - w - \phi\frac{c_u(p_l)}{2} - (1-\phi)\frac{c_u(p_h)}{2} \\
&\quad - \frac{1}{2}\big(\phi^2 p_l + (1-\phi^2)p_h - w - \phi^2 c_u(p_l) - (1-\phi)(c_u(p_h) - \phi c_o)\big) \\
&= \frac{1}{2}\big(\phi p_l + (1-\phi)p_h - w - \phi(1-\phi)(c_u(p_h) + c_o)\big), \\
\Delta\hat{\Lambda}_5 - \Delta\hat{\Omega}_5 &= \phi p_l + (1-\phi)p_h - \frac{r}{2} - \frac{w}{2} - \phi\frac{c_u(p_l)}{2} - (1-\phi)\frac{c_u(p_h)}{2} + c_u(r) \\
&\quad - \frac{1}{2}\big(\phi^2 p_l + (1-\phi^2)p_h - r - \phi^2 c_u(p_l) \\
&\quad - (1-\phi)c_u(p_h) - c_o(1 - \phi(1-\phi)) + c_u(r)\big) \\
&= \frac{1}{2}\big(\phi p_l + (1-\phi)p_h - w - \phi(1-\phi)(c_u(p_h) + c_o)\big),
\end{aligned}$$

and

$$\hat{\Lambda}_5(p) - \hat{\Omega}_5(p) = \Delta\hat{\Lambda}_5 - \Delta\hat{\Omega}_5 < 0, \quad \forall\phi \in \left[c(p_l), \frac{c(p_h)}{c(r)}\right].$$

We can conclude

$$\hat{\delta}_{A5}(\hat{f} = 1) < 0, \quad \forall\phi \in \left[c(p_l), \frac{c(p_h)}{c(r)}\right]. \tag{A.11}$$

3. Best response $\hat{f}_5^*$ Given the results of (A.10) and (A.11) the best response of retailer A in Case 5 is

$$\hat{f}_5^* = 0, \quad \forall \hat{f}_B \in [c(r), 1].$$

Case 6 $f \in [c(r), 1]$ $\phi \in \left[0, \frac{c(p_l)}{c(r)}\right]$

1. $\hat{\delta}_{A6}(\hat{f}_B = 0) = -\alpha\Delta\Upsilon + \beta_{1t}\hat{\Lambda}_6(p) + \beta_2\Delta\hat{\Lambda}_6$

$$\begin{aligned}
\hat{\Lambda}_6(p) &= \phi p_l + (1-\phi)p_h - w - (1-\phi)\frac{c_u(p_h)}{2} \\
&= \frac{1}{2}\big(\phi p_l + (1-\phi)p_h - w - g + \phi c_u(p_l)\big), \\
\Delta\hat{\Lambda}_6 &= \phi p_l + (1-\phi)p_h - \frac{r}{2} - \frac{w}{2} - (1-\phi)\frac{c_u(p_h)}{2} + \frac{c_u(r)}{2} \\
&= \frac{1}{2}(\phi p_l + (1-\phi)p_h - w + \phi c_u(p_l)),
\end{aligned}$$

$$\hat{\Lambda}_6(p) < 0, \quad \forall \phi \in \left[0, \frac{c(p_l)}{c(r)}\right],$$
$$\Delta\hat{\Lambda}_6 > 0, \quad \forall \phi \in \left[0, \frac{c(p_l)}{c(r)}\right].$$

Given that $\hat{\Lambda}_6(p) > \Delta\hat{\Lambda}_6$ and $\beta_{1t} > \beta_2$, we can summarize

$$\hat{\delta}_{A6}(\hat{f}_B = 0) < 0, \quad \forall \phi \in \left[0, \frac{c(p_l)}{c(r)}\right]. \tag{A.12}$$

2. $\hat{\delta}_{A6}(\hat{f}_B = 1) = -\alpha\Delta\Upsilon + \beta_{1t}\big(\hat{\Lambda}_6(p) - \hat{\Omega}_6(p)\big) + \beta_2\big(\Delta\hat{\Lambda}_6 - \Delta\hat{\Omega}_6\big)$, where

$$\begin{aligned}\hat{\Lambda}_6(p) - \hat{\Omega}_6(p) &= \phi p_l + (1-\phi)p_h - w - (1-\phi)\frac{c_u(p_h)}{2}\\ &\quad -\frac{1}{2}\big(\phi^2 p_l + (1-\phi^2)p_h - w - (1-\phi)c_u(p_h) + \phi c_o\big),\\ \Delta\hat{\Lambda}_6 - \Delta\hat{\Omega}_6 &= \phi p_l + (1-\phi)p_h - \frac{r}{2} - \frac{w}{2} - (1-\phi)\frac{c_u(p_h)}{2} + \frac{c_u(r)}{2}\\ &\quad -\frac{1}{2}\big(\phi^2 p_l + (1-\phi^2)p_h - r - (1-\phi)c_u(p_h) + \phi c_o + c_u(r)\big).\end{aligned}$$

The best response function in Case 6 is convex in ϕ as given by

$$\frac{\partial^2 \hat{\delta}_{A6}(\hat{f}_B = 1)}{\partial \phi^2} = (\beta_{1t} + \beta_2)(p_h - p_l) > 0.$$

Consequently, showing that the best response is negative can be achieved by showing that the best response is negative at either bound $\phi = 0$ and $\phi = 1$. Applying the transformations from (A.1), the best response at the bounds is given by

$$\hat{\delta}_{A6}\big(\hat{f} = 1, \phi = 0\big) = -\alpha(r - p_h) + \frac{\beta_{1t}}{2}(p_h - w) + \frac{\beta_2}{2}(p_h - w),$$

which is negative $\forall \alpha > (\beta_{1t} + \beta_2)\frac{p_h - w}{r - p_h}$ which is fulfilled.

$$\hat{\delta}_{A6}\big(\hat{f} = 1, \phi = 1\big) = -\alpha(r - p_l) - \left(\frac{\beta_{1t}}{2} + \frac{\beta_2}{2}\right)\big(-(p_l - w) + c_o\big),$$

with $\left(\frac{\beta_{1t}}{2} + \frac{\beta_2}{2}\right)\big(-(p_l - w) + c_o\big) > 0$.

We can conclude

$$\hat{\delta}_{A6}\big(\hat{f} = 1\big) < 0, \quad \forall \phi \in \left[0, \frac{c(p_l)}{c(r)}\right]. \tag{A.13}$$

3. Best response $\hat{f}_6^*$ From (A.12) and (A.13) we see that the best response of retailer A in Case 6 is given by

$$\hat{f}_6^* = 0, \quad \forall f \in [c(r), 1].$$

Case 7 $f \in [c(p_h), c(r)]$ $\phi \in \left[0, \frac{c(p_l)}{c(p_h)}\right]$

1. $\hat{\delta}_{A7}(\hat{f}_B = 0) = -\alpha\Delta\Upsilon + \beta_{1t}\hat{\Lambda}_7(p) + \beta_2\Delta\hat{\Lambda}_7$, where

$$\begin{aligned}
\hat{\Lambda}_7(p) &= \phi p_l + (1-\phi)p_h - w - (1-\phi)\frac{c_u(p_h)}{2} \\
&= \frac{1}{2}(\phi p_l + (1-\phi)p_h - w - g + \phi c_u(p_l)), \\
\Delta\hat{\Lambda}_7 &= \phi p_l + (1-\phi)p_h - \frac{r}{2} - \frac{w}{2} - (1-\phi)\frac{c_u(p_h)}{2} \\
&= \frac{1}{2}(\phi p_l + (1-\phi)p_h - r - g + \phi c_u(p_l)),
\end{aligned}$$

with

$$\begin{aligned}
\hat{\Lambda}(p) &< 0, \quad \forall\phi \in \left[0, \frac{c(p_l)}{c(p_h)}\right], \\
\Delta\hat{\Lambda}_7 &< 0, \quad \forall\phi \in \left[0, \frac{c(p_l)}{c(p_h)}\right],
\end{aligned}$$

$$\delta_{A7}(\hat{f}_B = 0) < 0, \quad \phi \in \left[0, \frac{c(p_l)}{c(p_h)}\right]. \tag{A.14}$$

2. $\hat{\delta}_{A7}(\hat{f}_B = c(r)) = -\alpha\Delta\Upsilon + \beta_{1t}(\hat{\Lambda}_7(p) - c(r)\hat{\Omega}_7(p)) + \beta_2(\Delta\hat{\Lambda}_7 - c(r)\Delta\hat{\Omega}_7)$, where

$$\begin{aligned}
\hat{\Lambda}_7(p) - c(r)\hat{\Omega}_7(p) &= \phi p_l + (1-\phi)p_h - w - (1-\phi)\frac{c_u(p_h)}{2} \\
&\quad - \frac{c(r)}{2}(\phi^2 p_l + (1-\phi^2)p_h - w - (1-\phi)(c_u(p_h) + \phi c_o)), \\
\Delta\hat{\Lambda}_7 - c(r)\Delta\hat{\Omega}_7 &= \phi p_l + (1-\phi)p_h - \frac{r}{2} - \frac{w}{2} - (1-\phi)\frac{c_u(p_h)}{2} \\
&\quad - \frac{c(r)}{2}(\phi^2 p_l + (1-\phi^2)p_h - r - (1-\phi)(c_u(p_h) + c_o)).
\end{aligned}$$

The best response function in Case 7 is convex in ϕ as given by

$$\frac{\partial^2\hat{\delta}_{A7}(\hat{f}_B = c(r))}{\partial\phi^2} = c(r)(\beta_{1t} + \beta_2)(p_h - p_l) > 0.$$

Consequently, showing that the best response is negative can be achieved by showing that the best response is negative at either bound $\phi = 0$ and $\phi = 1$. Applying the transformations from (A.1), the best response at the bounds is given by

$$\hat{\delta}_{A7}\big(\hat{f} = c(r), \phi = 0\big) = -\alpha(r - p_h) - \frac{\beta_{1t}}{2}\big(-(p_h - w) + g(1 - c(r))\big) + \frac{\beta_2}{2}(p_h - w),$$

with $\alpha(r - p_h) > \frac{\beta_2}{2}(p_h - w)$ and $\frac{\beta_{1t}}{2}\big(-(p_h - w) + g(1 - c(r))\big) > 0, \forall p_h - w < g(1 - c(r))$ we find that $\hat{\delta}_{A7}\big(\hat{f} = c(r), \phi = 0\big) < 0$.

$$\hat{\delta}_{A7}\big(\hat{f}{=}c(r), \phi{=}1\big) = -\alpha(r - p_l) - \beta_{1t}\Big(-(p_l - w) + \frac{c(r)}{2}(p_l - w + c_o)\Big) - \frac{\beta_2}{2}(r - p_l)(1 - c(r)) + \frac{\beta_2}{2}(p_l - w),$$

with $\alpha(r - p_l) > \frac{\beta_2}{2}(p_l - w)$ and $\beta_{1t}\big(-(p_l - w) + \frac{c(r)}{2}(p_l - w + c_o)\big) > 0$, $\forall p_l - w < g(1 - c(p_h))$ we find that $\hat{\delta}_{A7}\big(\hat{f} = c(r), \phi = 1\big) < 0$.
We can conclude

$$\hat{\delta}_{A7}\big(\hat{f} = c(r)\big) < 0, \quad \forall \phi \in \left[0, \frac{c(p_l)}{c(p_h)}\right]. \tag{A.15}$$

3. Best response $\hat{f}_7^*$ The best response of retailer A in Case 7, given (A.14) and (A.15),

$$\hat{f}_7^* = 0, \quad \forall \hat{f}_B \in [c(p_h), c(r)].$$

Case 8 $f \in [0, c(p_h)]$ and $\phi \in [0, 1]$

1. $\hat{\delta}_{A8}(\hat{f}_B = 0) = -\alpha \Delta \Upsilon + \beta_{1t} \hat{\Lambda}_8(p) + \beta_2 \Delta \hat{\Lambda}_8$, where

$$\hat{\Lambda}_8(p) = \phi p_l + (1 - \phi) p_h - w,$$
$$\Delta \hat{\Lambda}_8 = \phi p_l + (1 - \phi) p_h - \frac{w}{2} - \frac{r}{2},$$

$$\hat{\Lambda}(p) > 0, \quad \forall \phi \in [0, 1],$$
$$\Delta \hat{\Lambda}_8 < 0, \quad \forall \phi \in [0, 1].$$

And the best response is positive if

$$\beta_{1t} > \tau = \frac{\alpha \Delta \Upsilon - \beta_2 \Delta \Lambda}{\Lambda(p)},$$

which we denoted in the proposition by τ. Then we can summarize

$$\delta_{A8}(\hat{f}_B = 0) \begin{cases} > 0 & \text{if } \beta_{1t} > \tau, \\ < 0 & \text{if } \beta_{1t} < \tau. \end{cases} \tag{A.16}$$

2. $\hat{\delta}_{A8}(\hat{f}_B = c(p_h)) = -\alpha \Delta \Upsilon + \beta_{1t}\big(\hat{\Lambda}_8(p) - c(p_h)\hat{\Omega}_8(p)\big) + \beta_2\big(\Delta\hat{\Lambda}_8 - c(p_h)\Delta\hat{\Omega}_8\big)$, where

$$\begin{aligned} \hat{\Lambda}_8(p) - c(p_h)\hat{\Omega}_8(p) &= \phi p_l + (1-\phi)p_h - w \\ &\quad - \frac{c(p_h)}{2}\big(\phi^2 p_l + (1-\phi^2)(p_h - w + c_o)\big), \\ \Delta\hat{\Lambda}_8 - c(p_h)\Delta\hat{\Omega}_8 &= \phi p_l + (1-\phi)p_h - \frac{r}{2} - \frac{w}{2} \\ &\quad - \frac{c(p_h)}{2}\big(\phi^2 p_l + (1-\phi^2)p_h - r\big). \end{aligned}$$

The best response function in Case 8 is convex in ϕ as given by

$$\frac{\partial^2 \hat{\delta}_{A8}(\hat{f}_B = c(p_h))}{\partial \phi^2} = c(p_h)(\beta_{1t} + \beta_2)(p_h - p_l) > 0.$$

Consequently, showing that the best response is negative can be achieved by showing that the best response is negative at either bound $\phi = 0$ and $\phi = 1$. Applying the transformations from (A.1), the best response at the bounds is given by

$$\begin{aligned} \hat{\delta}_{A8}\big(\hat{f} = c(p_h), \phi = 0\big) &= -\alpha(r - p_h) - \beta_{1t}\left(-(p_h - w) + \frac{c(p_h)}{2}(p_h - w + c_o)\right) \\ &\quad - \frac{\beta_2}{2}(r - p_h)(1 - c(p_h)) + \frac{\beta_2}{2}(p_h - w), \end{aligned}$$

with $\alpha(r - p_h) > \frac{\beta_2}{2}(p_h - w)$ and $\beta_{1t}\big(-(p_h - w) + \frac{c(p_h)}{2}(p_h - w + c_o)\big) > 0$, $\forall\, p_l - w < g(1 - c(p_h))$ we find that $\hat{\delta}_{A8}\big(\hat{f} = c(p_h), \phi = 0\big) < 0$.

$$\begin{aligned} \hat{\delta}_{A8}\big(\hat{f} = c(p_h), \phi = 1\big) &= -\alpha(r - p_l) - \beta_{1t}\left(-(p_l - w) + \frac{c(p_h)}{2}(p_l - w + c_o)\right) \\ &\quad - \frac{\beta_2}{2}(r - p_l)(1 - c(p_h)) + \frac{\beta_2}{2}(p_l - w). \end{aligned}$$

We can conclude

$$\hat{\delta}_{A8}\big(\hat{f} = c(p_h)\big) < 0. \tag{A.17}$$

3. Best response $\hat{f}_8^*$ The best response in case 8 depends according to (A.16) on the size of the stockpiling segment β_{1t}. In case the stockpiling segment is below its critical size, i.e., $\beta_{1t} < \tau$, retailer A's best response is negative both at the lower bound and the upper bound of $f_B \in [0, c(p_h)]$ and consequently his best response is to play the regular price. To the contrary, if the stockpiling segment is larger than its critical size, i.e., $\beta_{1t} > \tau$ retailer A plays strategically and his best response is a mixed one as described by the proposition.

A.2 Best Response for the Scenario Information Sharing

Proposition A.2. *Under information sharing, the best response $\check{\delta}_{Ak}(\check{f}_B, \beta_{1t})$ of retailer A depends on the size of the stockpiling segment β_{1t} and the competitor's promotion frequency $\check{f}_B$. The critical size for the stockpiling segment τ is derived as*

$$\begin{aligned}\tau &= \frac{\alpha \Delta \Upsilon - \beta_2 \Delta \Lambda}{\Lambda(p)} \\ &= \frac{\alpha(r - \phi p_l - (1-\phi)p_h) - \beta_2(\phi p_l + (1-\phi)p_h - \frac{r}{2} - \frac{w}{2})}{\phi p_l + (1-\phi)p_h - w},\end{aligned}$$

and we can describe the following two cases:

Case i: *If $\beta_{1t} < \tau$, then $\check{\delta}_{Ak}(\check{f}_B, \beta_{1t}) < 0 \; \forall \check{f}_B \in [0,1]$ and retailer A's best response is to play his pure strategy "regular" no matter how retailer B plays: He has a dominant best response.*

$$\check{f}_A^* = 0.$$

Case ii: *If $\beta_{1t} > \tau$, Retailer A's best response is to play strategically and mix between his pure strategies "regular" and "promotion" with probability*

$$\check{f}_A^* = \begin{cases} 1 & \text{if } \check{f}_B \in [0, \check{f}_B^\dagger), \\ [0,1] & \text{if } \check{f}_B = \check{f}_B^\dagger, \\ 0 & \text{if } \check{f}_B \in (\check{f}_B^\dagger, 1], \end{cases}$$

with

$$\check{f}_B^\dagger = \frac{-\alpha \Delta \Upsilon + \beta_{1t} \check{\Lambda}(p) + \beta_2 \Delta \check{\Lambda}}{\beta_{1t} \check{\Omega}_k(p) + \beta_2 \Delta \check{\Omega}_k}.$$

Proof. We use the three step procedure as described in Sect. 4.2.2.2 for each of the $k = 3$ cases:

Case 1 $\check{f}_B \in [0,1], \phi \in [c(p_h),1]$

1. $\check{\delta}_{A1}(\check{f}_B = 0) = -\alpha\Delta\Upsilon + \beta_1\check{\Lambda}(p) + \beta_2\Delta\check{\Lambda}$

$$\check{\Lambda}_1(p) = \phi p_l + (1-\phi)p_h - w,$$
$$\Delta\check{\Lambda}_1 = \phi p_l + (1-\phi)p_h - \frac{r}{2} - \frac{w}{2},$$

where

$$\check{\Lambda}_1(p) > 0, \quad \forall\phi \in [0,1],$$
$$\Delta\check{\Lambda}_1 < 0, \quad \forall\phi \in [0,1],$$

and the best response is positive, if

$$\beta_{1t} > \tau = \frac{\alpha\Upsilon - \beta_2\Delta\check{\Lambda}_1}{\check{\Lambda}_1(p)}$$
$$= \frac{\left(\alpha + \frac{\beta_2}{2}\right)\left(r - \phi p_l - (1-\phi)p_h\right) - \frac{\beta_2}{2}\left(\phi p_l + (1-\phi)p_h - w\right)}{\phi p_l + (1-\phi)p_h - w},$$

which we denote by τ. Then we can summarize

$$\check{\delta}_{A1}(\hat{f}_B = 0)\begin{cases} > 0 & \text{if} \quad \beta_{1t} > \tau, \\ < 0 & \text{if} \quad \beta_{1t} > \tau. \end{cases} \tag{A.18}$$

Observe that this condition holds true for cases 1–3. We will therefore omit this first step in the further cases and refer to above proof.

2. $\check{\delta}_{A1}(\check{f}_B = f_{o1}) - \alpha\Delta\Upsilon + \beta_{1t}\left(\check{\Lambda}(p) - \check{\Omega}_1(p)\right) + \beta_2\left(\Delta\check{\Lambda} - \Delta\check{\Omega}_1\right)$ with $f_{o1} = 1$. Observe that due to $\check{f}_{oB} = 1$,

$$\check{\Lambda}(p) - \check{\Omega}_k(p) = \Delta\check{\Lambda} - \Delta\check{\Omega}_k,$$

which holds true for all cases $k = 1, \ldots, 3$ and we substitute the term by $\check{X}_k$

$$\check{X}_1 = \phi p_l + (1-\phi)p_h - w$$
$$- \frac{1}{2}(\phi^2 p_l + (1-\phi^2)p_h - w + (1-\phi)(\phi c_u(p_l) + (1-\phi)c_u(p_h))), \tag{A.19}$$

where

$$\check{X}_1 \begin{cases} < 0 \; \forall\phi \in \left[c(p_h), \frac{g}{c_u(p_l)}\right], \\ > 0 \; \forall\phi \in \left[\frac{g}{c_u(p_l)}, 1\right]. \end{cases}$$

$\forall\phi < \frac{g}{c_u(p_l)}$ $\check{X}_1$ is negative and consequently the best response is negative.
$\forall\phi > \frac{g}{c_u(p_l)}$ $\check{X}_1$ is positive and we attain a bound for the size of the stockpiling segment at

$$\beta_{1t} < \frac{\alpha\Upsilon - \frac{\beta_2}{2}(\phi c_u(p_l) - g)}{\phi c_u(p_l) - g},$$

which is always fulfilled due to a small denominator and consequently a large fraction which will not be exceeded by the size of the stockpiling segment β_{1t}. Consequently, we can conclude that

$$\check{\delta}_{A1}(\check{f}_B = 1) < 0, \quad \forall\phi \in [c(p_h), 1]. \tag{A.20}$$

3. Best response $\check{f}^*_{A1}$
   Combining (A.18) and (A.20), we find that retailer A has a dominant best response if $\beta_{1t} < \tau$:

$$\forall\beta_{1t} < \tau, \quad f^*_{A1} = 0.$$

   To the contrary, he plays strategically if the size of the stockpiling segment is large enough: if $\beta_{1t} > \tau$, his best response is

$$\forall\beta_{1t} > \tau, \quad \check{f}^*_{A1} = \begin{cases} 1 & \text{if } \check{f}_B \in [0, \check{f}^\dagger_{B1}), \\ [0,1] & \text{if } \check{f}_B = \check{f}^\dagger_{B1}, \\ 0 & \text{if } \check{f}_B \in (\check{f}^\dagger_{B1}, 1], \end{cases}$$

   with

$$\check{f}^\dagger_{B1} = \frac{-\alpha\Upsilon + \beta_{1t}\check{\Lambda}(p) + \beta_2\Delta\check{\Lambda}}{\beta_1\check{\Omega}_1(p) + \beta_2\Delta\check{\Omega}_1}.$$

Case 2 $\check{f}_B \in [0,1], \phi \in [c(p_l), c(p_h)]$

1. $\check{\delta}_{A2}(\check{f}_B = 0)$ See Case 1.
2. $\check{\delta}_{A2}(\check{f}_B = f_{o2})$ with $f_{o2} = 1$

$$\begin{aligned}\check{X}_2 &= \phi p_l + (1-\phi)p_h - w - \frac{1}{2}(\phi^2 p_l + (1-\phi^2)p_h - w + \phi(1-\phi)(c_u(p_l) + c_o)) \\ &= \frac{1}{2}(p_h - w - \phi(p_h - p_l + (1-\phi)(c_u(p_h) + c_o))),\end{aligned}$$

   where

$$\check{X}_2 < 0, \quad \forall\phi \in [c(p_l), c(p_h)].$$

Consequently, we can conclude that

$$\check{\delta}_{A2}(\check{f}_B = 1) < 0, \quad \forall \phi \in [c(p_l), c(p_h)]. \tag{A.21}$$

3. Best response $\check{f}^*_{A2}$
   Combining (A.18) and (A.21), we find that retailer A has a dominant best response if $\beta_{1t} < \tau$:
   $$\forall \beta_{1t} < \tau, \quad f^*_{A2} = 0.$$
   To the contrary, he plays strategically if the size of the stockpiling segment is large enough: if $\beta_{1t} > \tau$, his best response is
   $$\forall \beta_{1t} > \tau, \quad \check{f}^*_{A2} = \begin{cases} 1 & \text{if } \check{f}_B \in [0, \check{f}^\dagger_{B2}), \\ [0,1] & \text{if } \check{f}_B = \check{f}^\dagger_{B2}, \\ 0 & \text{if } \check{f}_B \in (\check{f}^\dagger_{B2}, 1], \end{cases}$$
   with
   $$\check{f}^\dagger_{B2} = \frac{-\alpha\Upsilon + \beta_{1t}\check{\Lambda}(p) + \beta_2\Delta\check{\Lambda}}{\beta_1\check{\Omega}_2(p) + \beta_2\Delta\check{\Omega}_2}.$$

Case 3 $\check{f}_B \in [0,1], \phi \in [0, c(p_l)]$

1. $\check{\delta}_{A3}(\check{f}_B = 0)$ See Case 1.
2. $\check{\delta}_{A3}(\check{f}_B = f_{o2})$ with $f_{o3} = 1$

$$\begin{aligned}\check{X}_3 &= \phi p_l + (1-\phi)p_h - w - \frac{1}{2}\left(\phi^2 p_l + (1-\phi^2)p_h - w + \phi(1-\phi)c_o\right) \\ &= \frac{1}{2}\left(\phi\,(2-\phi)\,p_l + (1-\phi)^2 - w - \phi c_o\right),\end{aligned}$$

where

$$\check{X}_3 < 0, \quad \forall \phi \in (0, c(p_l)]$$

Consequently, we can conclude that

$$\check{\delta}_{A3}(\check{f}_B = 1) < 0, \quad \forall \phi \in [0, c(p_l)]. \tag{A.22}$$

3. Best response $\check{f}^*_{A3}$
   Combining (A.18) and (A.22), we find that retailer A has a dominant best response if $\beta_{1t} < \tau$:
   $$\forall \beta_{1t} < \tau, \quad f^*_{A3} = 0.$$

To the contrary, he plays strategically if the size of the stockpiling segment is large enough: if $\beta_{1t} > \tau$, his best response is

$$\forall \beta_{1t} > \tau, \quad f_{A3}^{*} = \begin{cases} 1 & \text{if } \check{f}_B \in [0, f_{B3}^{\dagger}), \\ [0, 1] & \text{if } \check{f}_B = f_{B3}^{\dagger}, \\ 0 & \text{if } \check{f}_B \in (f_{B3}^{\dagger}, 1], \end{cases}$$

with

$$f_{B3}^{\dagger} = \frac{-\alpha \Upsilon + \beta_{1t} \check{\Lambda}(p) + \beta_2 \Delta \check{\Lambda}}{\beta_1 \check{\Omega}_3(p) + \beta_2 \Delta \check{\Omega}_3}.$$

□

References

Ailawadi, K. L., Gedenk, K., Lutzky, C., & Neslin, S. A. (2007). Decomposition of the sales impact of promotion-induced stockpiling. *Journal of Marketing Research*, *44*, 450–567.

Arminger, G. (2003). Die Revolution in der automatischen Disposition.

Assuncao, J. L., & Meyer, R. L. (1993). The rational effect of price promotions and sales on consumption. *Management Science*, *39*(5), 517–535.

Aviv, Y. (2001). The effect of collaborative forecasting on supply chain performance. *Management Science*, *47*(10), 1326–1343.

Banks, J., & Moorthy, S. (1999). A model of price promotions with consumer search. *International Journal of Industrial Organization*, *17*(17), 371–398.

Bell, D. R., & Botzug, Y. (2004). *The effect of inventory on purchase incidence: Empirical analysis of opposing foreces of storage and consumption* (Working paper).

Bell, D. R., & Hilber, C. A. L. (2006). An empirical test of the theory of sales: Do household storage constraints affect consumer and store behavior? *Quantitative Marketing and Economics*, *4*, 87–117.

Bell, D. R., & Lattin, J. M. (1998). Shopping behaviour and consumer preference for store price format: Why large basket shoppers prefer edlp. *Marketing Science*, *17*(1), 66–88.

Bell, D. R., Chiang, J., & Padmanabhan, V. (1999). The decompostion of promotional response: An empirical generalization. *Marketing Science*, *18*(4), 504–526.

Bell, D. R., Iyer, G., & Padmanabhan, V. (2002). Price competition under stockpiling and flexible consumption. *Journal of Marketing Research*, *49*, 292–303.

Ben-Akiva, M., & Lerman, S. R. (1985). *Discrete Choice Analysis: Theory and Application to Travel Demand*. Boston: The MIT Press.

Bernstein, F., & Federgruen, A. (2004). A general equilibrium model for industries with price and service competition. *Operations Research*, *52*(6), 868–886.

Blattberg, R. C., & Neslin, S. A. (1990). *Sales Promotion – Concepts, Methods and Strategies*. Upper Saddle River, NJ: Prentice Hall.

Blattberg, R. C., Eppen, G. D., & Liebermann, J. (1981). A theoretical and empirical evaluation of price deals for consumer nondurables. *Journal of Marketing*, *45*, 116–129.

Boizot, C., Robin, J.-M., & Visser, M. (2001). The demand for food products: An analysis of interpurchase times and purchased quantities. *Economic Journal*, *111*(470), 391–419.

Bucklin, R. E., & Gupta, S. (1992). Brand choice, purchase incidence and segmentation: An integrated modelling approach. *Journal of Markting Research*, *29*, 201–215.

Bucklin, R. E., Gupta, S., & Siddarth, S. (1998). Modelling the effect of purchase quantity on consumer choice of product assortment. *Journal of Forecasting*, *17*, 281–301.

Bunn, C., & Banks, J. (2004). Promotions: Adding value or driving sales? *Admap*, (451), 20–23.

Cachon, G. P., & Fisher, M. (2000). Supply chain inventory management and the value of shared information. *Management Science*, *46*(8), 1032–1048.

Cachon, G. P., & Kök, A. G. (2007). Category management and coordination in retail assortment planning in the presence of basket shopping consumers. *Management Science*, *53*(6), 934–951.

Cachon, G. P., & Lariviere, M. A. (2001). Contracting to assure supply: How to share demand forecasts in a supply chain. *Management Science, 47*(5), 629–646.

Cachon, G. P., & Netessine, S. (2004). Game theory in supply chain analysis. In D. Simchi-Levi, S. D. Wu, & Z.-J. Shen (Eds.), *Handbook of Quantitative Supply Chain Analysis: Modeling in the eBusiness Era*. Dordrecht: Kluwer.

Cachon, G. P., Randall, T., & Schmidt, G. (2007). In search of the bullwhip effect. *Management Science, 9*(4), 457–479.

Chandon, P., & Wansink, B. (2002). When are stockpiled products consumed faster? *Journal of Marketing Research, 39*(3), 321–335.

Chen, F. (2003). Information sharing and supply chain coordination. In S. Graves & T. de Kok (Eds.), *Handbooks in Operations Research and Management Science*. Amsterdam: North Holland.

Chen, F., & Yu, B. (2005). Quantifying the value of leadtime information in a single-location inventory system. *Manufacturing & Service Operations Management, 7*(2), 144–151.

Chen, Y., Narasimhan, C., & Zhang, Z. J. (2001). Research note: Consumer heterogeneity and competitive price-matching guarantees. *Marketing Science, 20*(3), 300–314.

Cheng, F., & Sethi, S. P. (1999). A periodic review inventory model with demand influenced by promotion decisions. *Management Science, 45*(11), 1510–1523.

Chintagunta, P. K. (1993). Investigating purchase incidence, brand choice and purchase quantity decisions of households. *Marketing Science, 12*(2), 184–208.

Croson, R., & Donohue, K. (2005). Upstream versus downstream information and its impact on the bullwhip effect. *System Dynamics Review, 21*(3), 249–260.

ECR (2004). *ECR Studie 2004 – Umsetzung von innovativen Technologien und ECR-Prozessen in der deutschsprachigen Konsumgüterindustrie und im Handel*. Köln.

Edgeworth, F. Y. (1888). The mathematical theory of banking. *Journal of the Royal Statistical Society, 51*, 113–127.

Elmaghraby, W., & Keskinocak, P. (2003). Dynamic pricing in the presence of inventory considerations: Research overview, current practices and future directions. *Management Science, 49*(10), 1287–1309.

Farmer, A. (1994). Information sharing with capacity uncertainty: The case of coffee. *Canadian Journal of Economics, 27*(2), 415–432.

Fleischmann, M., Hall, J. M., & Pyke, D. F. (2004). Smart pricing. *MIT Sloan Management Review, Winter 2004*, 9–13.

Fox, E. J., Montgomery, A. L., & Lodish, L. M. (2004). Consumer shopping and spending across retail formats. *Journal of Business, 77*(2), 25–60.

Fudenberg, D., & Tirole, J. (1998). *Game Theory* (6th ed.). Cambridge, MA: The MIT Press.

Gal-Or, E. (1985). Information sharing in oligopoly. *Econometrica, 53*(2), 329–343.

Gal-Or, E. (1986). Information transmission: Cournot and Bertrand equilibrium. *Review of Economic Studies, 53*, 85–92.

Gavirneni, S., Kapuscinski, R., & Tayur, S. (1999). Value of information in capacitated supply chains. *Management Science, 45*(1), 16–24.

GCI & Capgemini (2008). Future Supply Chain 2016: Serving consumers in a sustainable way.

Guadani, P. M., & Little, J. D. C. (1983). A logit model of brand choice calibrated on scanner data. *Marketing Science, 2*(3), 203–238.

Gupta, S. (1988). Impact of sales promotions on when, what and how much to buy. *Journal of Marketing Research, 25*, 342–355.

Hall, J. M., Kopalle, P. K., & Krishna, A. (2003). *A category management model of retailer dynamic pricing and ordering decisions: Normative and emprical analysis* (Working paper).

Harms, T. (2004). *Global pricing trends* (Technical report). Ernst & Young.

Hendel, I., & Nevo, A. (2003). The post-promotion dip puzzle: What do the data have to say? *Quantitative Marketing & Economics, 1*(4), 409–424.

Hendel, I., & Nevo, A. (2004). Intertemporal substitution and storable products. *Journal of the European Economic Association, 2*(2/3), 536–547.

Homburg, C., Hoyer, W. D., & Fassnacht, M. (2002). Service orientation of a retailer's business strategy: Dimensions, antecedents, and performance outcomes. *Journal of Marketing*, *66*(4), 86–101.
Hopp, A. (2005). *Index über den erwarteten Wettbewerbsdruck* (Presentation).
Huchzermeier, A., Iyer, A., & Freiheit, J. (2002). The supply chain impact of smart customers in a promotional environment. *Manufacturing & Service Operations Management*, *4*(3), 228–240.
Huchzermeier, A., Burkhardt, D., & Artmann, C. (2005). *Kraft Food Deutschland Gmbh: Pioneering ECR* (Case study). WHU – Otto Beisheim School of Management.
Iyer, A. V., & Ye, J. (2000). Assessing the value of information sharing in a promotional environment. *Management Science*, *2*(2), 128–143.
Jain, A., & Moinzadeh, K. (2005). A supply chain model with reverse information exchange. *Manufacturing & Service Operations Management*, *7*(4), 360–378.
Khouja, M. (1999). The single-period news-vendor problem: Literature review and suggestions for future research. *Omega*, *27*(5), 537–553.
Khouja, M., & Robbins, S. S. (2003). Linking advertising and quantity decisions in the single-period inventory model. *International Journal of Production Economics*, *86*, 93–105.
Krishna, A. (1994). The impact of dealing patterns on purchase behavior. *Marketing Science*, *13*(4), 351–373.
Kulp, S. C., Lee, H. L., & Ofek, E. (2004). Manufacturer benefits from information integration with retail customers. *Management Science*, *50*(4), 431–444.
Kumar, V., & Leone, R. P. (1988). Measuring the effect of retail store promotions on brand and store substitiution. *Journal of Marketing Research*, *25*, 178–185.
Kurt Salmon Associates (1993). *Efficient Consumer Response – Enhancing Consumer Value in the Grocery Industry*. Washington, D.C.
Lach, S. (2002). Existence and persistence of price dispersion: An empirical analysis. *Review of Economics & Statistics*, *84*(3), 433–444.
Lal, R., & Rao, R. (1997). Supermarket competition: The case of every day low pricing. *Marketing Science*, *16*(1), 60–80.
Lal, R., Little, J. D. C., & Villas-Boas, J. M. (1996). A theory of forward buying, merchandising and trade deals. *Marketing Science*, *15*(1), 21–37.
Lee, H. L., & Whang, S. (2000). Information sharing in a supply chain. *International Journal of Technology Management*, *20*, 373–387.
Lee, H. L., Padmanabhan, V., & Whang, S. (1997). Information distortion in a supply chain: The bullwhip effect. *Management Science*, *43*(4), 546–558.
Lee, H. L., So, K. C., & Tang, C. S. (2000). The value of information sharing in a two-level supply chain. *Management Science*, *46*(5), 626–644.
Leszczyc, P. T. L. P., Sinha, A., & Timmermans, H. J. P. (2000). Consumer store choice dynamics: An analysis of the competitive market structure for grocery stores. *Journal of Retailing*, *73*(3), 323–345.
Li, L. (1985). Cournot oligopoly with information sharing. *Rand Journal of Economics*, *16*(4), 521–536.
Li, L. (2002). Information sharing in a supply chain with horizontal competition. *Management Science*, *48*(9), 1196–1212.
Metro Group (2006). *Metro-Handelslexikon 2006/2007 – Daten, Fakten und Adressen zum Handel in Deutschland, Europa und weltweit*. Düsseldorf.
Nahmias, S. (2001). *Production and operations analysis*. (4th ed.). New York: McGraw-Hill/Irwin.
Narasimhan, C. (1984). A price discrimination theory of coupons. *Marketing Science*, *3*(2), 128–147.
Narasimhan, C. (1988). Competitive promotional strategies. *Journal of Business*, *61*(4), 427–449.
Nash, J. (1951). Non-cooperative games. *Annals of Mathematics*, *54*(2), 286–295.
Nielsen, A. (2003). *The power of private label – A review of growth trends around the world* (Working paper).
Nijs, V. R., Dekimpe, M. G., Steenkamp, J.-B. E. M., & Hanssens, D. M. (2001). The category-demand effect of price promotions. *Marketing Science*, *20*(1), 1–22.

Novshek, W., & Sonnenschein, H. (1982). Fulfilled expectations cournot duopoly with information acquisition and release. *Bell Journal of Economics*, *13*, 214–218.

Pauwels, K. (2004). How dynamic consumer response, competitor response, company support, and company inertia shape long-term marketing effectiveness. *Marketing Science*, *23*(4), 596–610.

Petruzzi, N. C., & Dada, M. (1999). Pricing and the newsvendor problem: A review with extensions. *Operations Research*, *47*(2), 183–194.

Polman, P. (2004). Better shopping. In *ECR Europe 2004*, Brussels.

Raghunathan, S. (2001). Information sharing in a supply chain: A note on its value when demand is nonstationary. *Management Science*, *47*(4), 605–610.

Rajiv, S., Dutta, S., & Dhar, S. K. (2002). Asymmetric store positioning and promotional advertising strategies: Theory and evidence. *Marketing Science*, *21*(1), 74–96.

Raju, J. S., Srininvasan, V., & Lal, R. (1990). The effect of brand loyalty on competitive price promotional strategies. *Management Science*, *30*(3), 276–304.

Rao, R. C. (1991). Pricing and promotions in asymmetric duopolies. *Marketing Science*, *10*(2), 131–144.

Rao, R. C., Arjunji, R. V., & Murthi, B. P. S. (1995). Game theory and empirical generalizations concerning competitive promotions. *Marketing Science*, *14*(3), 89–100.

Rhee, H., & Bell, D. R. (2002). The inter-store mobility of supermarket shoppers. *Journal of Retailing*, *78*, 225–237.

Richards, T. (2005, July 24–27). *A nested logit model of strategic promotion*, 2005 Annual Meeting, American Agricultural Economics Association.

Rode, J. (2005). Metro verbessert kritische Promotions. *Lebensmittelzeitung*, (5).

Rudi, N., & Pyke, D. F. (2000). Teaching supply chain concepts with the newsboy model. In: M. E. Johnson & D. F. Pyke (Eds.), *Supply chain management: Innovations for education* (pp. 170–180). Miami: POMS.

Salop, S., & Stiglitz, J. E. (1982). The theory of sales: A simple model of equilibrium price dispersion with identical agents. *American Economic Review*, *72*, 1121–1130.

Schulz, H.-J. (2009). Lidl startet Werbekampagne in der Schweiz. *Lebensmittelzeitung*, *12*.

Seifert, D. (2002). *Collaborative Planning, Forecasting and Replenishment, How to create a supply chain advantage* (preprint version). Bonn: Galileo Press GmbH.

Shankar, V., & Bolton, R. N. (2004). An empriciral analysis of determinants of retailer pricing strategy. *Marketing Science*, *23*(1), 28–49.

Silver, E. A., Pyke, D. F., & Peterson, R. (1998). *Inventory management and production planning and scheduling* (3rd ed.). New York: Wiley.

Simester, D. (1997). Optimal promotion strategies: A demand-sided characterization. *Management Science*, *43*(2), 251–256.

Simon, H. (1995). *Preismanagement kompakt: Probleme und Methoden des modernen pricing*. Wiesbaden: Gabler.

Simon, H., & Fassnacht, M. (2009). *Preismanagement: Strategie – Analyse – Entscheidung – Umsetzung*. Wiesbaden: Gabler.

Sogomonian, A. G., & Tang, C. S. (1993). A modeling framework for coordinating promotion and production decisions within a firm. *Management Science*, *39*(2), 191–203.

Steckel, J. H., Gupta, S., & Banerji, A. (2004). Supply chain decision making: Will shorter cycle times and shared point-of-sale information necessarily help? *Management Science*, *50*(4), 458–464.

Steenkamp, J.-B. E. M., Nijs, V. R., Hanssens, D. M., & Dekimpe, M. G. (2005). Competitive reactions to advertising and promotion attacks. *Marketing Science*, *24*(1), 35–54.

Tang, C. S., & Yin, R. (2007). Joint ordering and pricing strategies for managing substitutable products. *Production and Operations Management*, *16*(1), 138–153.

Tang, C. S., Bell, D. R., & Ho, T.-H. (2001). Store choice and shopping behavior: How price format works. *California Management Review*, *43*(2), 46–74.

Terwiesch, C., Ren, Z. J., Ho, T. H., & Cohen, M. A. (2005). An empirical analysis of forecast sharing in the semiconductor equipment supply chain. *Management Science*, *51*(2), 208–220.

van Heerde, H. J., Leeflang, P. S. H., & Wittink, D. R. (2002). Flexible decomposition of price promotion effects using store-level scanner data (Working paper 02-107). Marketing Science Institute.

van Heerde, H. J., Gupta, S., & Wittink, D. R. (2003). Is 75% of the sales promotion bump due to brand switching? No, only 33% is. *Journal of Marketing Research*, *40*, 481–491.

van Heerde, H. J., Leeflang, P. S. H., & Wittink, D. R. (2004). Decomposing the sales promotion bump with store data. *Marketing Science*, *23*(3), 317–334.

Varian, H. R. (1980). A model of sales. *American Economic Review*, *70*, 651–659.

VICS (2004). Retail Event Collaboration – Business process guide.

Villas-Boas, J. M. (1995). Models of competitive price promotions: Some empirical evidence from the coffee and saltine crackers markets. *Journal of Economics and Management Strategy*, *4*, 85–107.

Vives, X. (1984). Duoploy information equilibrium: Cournot and Bertrand. *Journal of Economic Theory*, *34*, 71–94.

Vives, X. (1999). *Olibopoly pricing: Old ideas and new tools*. Cambridge, MA: The MIT Press.

Within, T. M. (1955). Inventory control and price theory. *Management Science*, *2*(1), 61–68.

Zhang, H. (2002). Vertical information exchange in a supply chain with duopoly retailers. *Production and Operations Management*, *11*(4), 531–546.

Zhao, X., & Xie, J. (2002). Forecasting errors and the value of information sharing in a supply chain. *International Journal of Production Research*, *40*(2), 311–335.

Breinigsville, PA USA
21 September 2010

245813BV00003B/10/P